THE
A... R'S
...OOK

...avis

...ress
...nks
...1992

Library of Congress Cataloging-in-Publication Data

Davis, T. Neil.
The aurora watcher's handbook / Neil Davis
p. cm.
Includes bibliographical references and index.
ISBN 0-912006-59-5 (alk. paper). -- ISBN 0-912006-60-9 (pbk : alk. paper)
1. Auroras. I. Title
QC971.D38 1992 91-43080
538'.786--cd20 CIP

 First printing.
International Standard Book Number: cloth 0-912006-59-5
paper 0-912006-60-9
Library of Congress Catalogue Number: 91-43080

Text and black and white photographs printed by
Thomson-Shore, Inc.
on 60# Glatfelter Thor Offset, recycled and acid-free.
Color plates printed by
Alan Lithograph Inc.
on 70# Paloma Matte Book, acid-free.

This publication was printed on acid-free paper that meets the minimum requirements of American National Standard for Information Sciences-Permanence for Paper for Printed Library Materials, ANSI Z39.48-1984.

Publication coordination, design and production by
Deborah Van Stone.

Cover photograph: Aurora of the southern hemisphere photographed from a NASA Shuttle in 1991. (NASA photograph courtesy of T. J. Hallinan and Don Lind.)

CONTENTS

PREFACE

No one really needs any instruction to enjoy the aurora (the northern lights). However, the more a person knows about this spectacular phenomenon, the more pleasurable the experience. Much is known about the aurora, but enough mystery remains to satisfy all who chose to delve into the topic.

In this handbook I have attempted to present material in such a way that a reader can quickly obtain information useful for observing and photographing the aurora. The first few pages give a brief explanation of the aurora and some practical hints for the observer and photographer. Section 4 is designed to give modest general familiarity with auroral phenomena, and the remainder of Part I gives more details.

Part II presents additional background that brings deeper understanding of auroral and related phenomena. This is the part of the book to be read when the aurora is hiding behind the clouds and a person has time for contemplation. Because Section 9 contains concepts perhaps unfamiliar, the section may require thoughtful reading. Some readers may choose to skip over it altogether, but any time expended here is likely to make one more comfortable with other material presented in Part II.

A reader need not assimilate the second part of the handbook before delving into Part III, a discussion of some of the unknowns. A particularly interesting unknown is the riddle of auroral sound.

Readers may notice a slight bias toward Alaska as an observing location. Alaska is one of the best places to observe the aurora, and it is easily accessible to those who might choose to incorporate aurora watching in their travel plans. But whether you watch this awesome natural spectacle from Alaska or elsewhere, have fun, and may the aurora always brighten you sky.

ACKNOWLEDGEMENTS

Contributing much to this work are the auroral scientists of the Geophysical Institute of the University of Alaska Fairbanks who read draft versions of the manuscript and suggested improvements. They are Dr. Syun-Ichi Akasofu, director of the institute; Dr. Thomas J. Hallinan; Neal B. Brown; and Daniel L. Osborne. Special thanks also go to Dr. Victor P. Hessler, Mr. Gus Lamprecht and Dr. David C. Fritts for allowing me to select auroral photographs from their fine collections, and also to Dr. Robert D. Hunsucker and Dr. Takeshi Ohtake for use of individual photographs. Drs. Don Lind (a former astronaut now at Utah State University) and Thomas J. Hallinan provided me with the excellent photographs acquired by NASA astronauts on the STS-39 Shuttle mission flown in April and May of 1991. More than special thanks go to Carla Helfferich, the managing editor of the University of Alaska Press, who both edits and writes scientific and other materials. My wife Rosemarie also helped with editing. The cartoons scattered here and there are the product of Patricia Davis Candler, an Alaska silversmith by profession who also entertains with her light-hearted sketches. It is also a pleasure to acknowledge Deborah Van Stone, the University of Alaska Press' manager, who aside from other work, did the typesetting and layout of the book. I and all persons mentioned in this acknowledgement are past or present members of the staff or former students of the University of Alaska Fairbanks, so the handbook truly is a product of this institution.

Part I
MATTERS OF IMMEDIATE CONCERN TO AURORA WATCHERS

Section 1
CAUSE OF THE AURORA (Briefly Explained)

A brilliant green aurora, its skirt hemmed in red, flashes across the sky. It pauses, then twirls about and steps out again to disappear beyond the tree-tops, and another follows along behind. The shivering man who stands in his driveway below has seen hundreds of auroras before, but this one, like all the rest, brings a tingling to his spine. He almost forgets that his young daughter is standing beside him, her head also tipped back in awe, until she asks, "Daddy, what are the northern lights, and what makes them happen?"

Recognizing that the girl is very young and deserving of a short but not misleading answer, the man says, "Well, they are actual lights very high up in the sky, and as you know, you can see them only at night. The northern lights—people also call it the aurora—are just like the neon lights that you see in store windows. Inside those neon lights are certain gases, and when electricity runs through those gases they glow in certain colors."

If the girl had been a little younger, this answer would have been all she needed. She would have walked back into the house entirely happy, and unaware of how much she really had been told. She had just learned that the northern lights—the aurora—is an actual source of light and that it is located high in the atmosphere. Like the light of the stars, the source is bright enough to be seen at night but not in the daytime, and like a neon-type sign it can be multi-colored (if filled with different kinds of

What he says...and what she hears.

gases). Furthermore, she heard a strictly correct statement about how auroral light is produced, for the mechanism is the same as in the neon light—electrical discharge in a gas of low density.

But then she asks, "How high is the aurora?" and her father has a ready answer.

"It's about 50 miles up over our heads. You know how far 50 miles is? No? Well that's about as far as Grandma's house. It takes us about an hour to drive over there, and that's how high the aurora is."

"Wow, that is high." Her father smiles inwardly because he knows his daughter has just absorbed a lesson yet to be learned by many adults who, having seen distant aurora meet the horizon, or having thought they saw it between them and a nearby hill, assume that the aurora comes down low, even perhaps touching the ground. It never does; the aurora is never lower than 40 miles (60 km), and it may extend upwards above the earth's surface for several hundred miles.

The father and daughter go back inside to warm up, now that the best of the aurora is gone. Then she says, "You told me that the northern lights are just like a neon sign. I don't know how a neon sign works. How does it work?"

"Well, you know that a neon sign is different from an ordinary light bulb. The light bulb has a little wire inside called a filament, and when electricity runs through the filament it gets so hot that it glows, first red, then yellowish and then white, like sunlight. The hotter the filament gets the brighter and more white the light. But neon signs give off light of just certain colors. The neon sign does not have a filament like the light bulb does. Instead, it is filled with a gas that glows, and the color of the light given off depends on the kind of gas inside. The aurora is like that, too. Its colors depend on the kinds of gas high up in the sky, in what we call the earth's high atmosphere. Up there, as here near the ground, the air is mostly nitrogen and oxygen."

The girl seems to understand that, but then she asks, "O.K., but *why* does the gas glow?" Hmm, she is persistent, just like her mother, thinks the father.

"The gas glows because electricity is running through it. When you hook up a neon tube to electrical wires it causes the electricity to run through the gas in the tube. The electricity is actually carried by tiny little pieces of matter called electrons. As they go through the tube they run into the gas in the tube and make it give off light. In an ordinary light bulb the filament glows, but in the neon tube, the gas everywhere inside the tube gives off the light. The gas is in little pieces called atoms and molecules that float around. When an electron hits one of the atoms or molecules a little burst of light is given off.[1] It's the same in the aurora. Electrons come down into the atmosphere from above and they hit the atoms and molecules and make them glow. If enough of the electrons come into the air high over our heads, all those flashes

1. A fluorescent light is a little different in that its light is given off when electrons flowing through the tube hit fluorescent material painted on the tube's inside walls.

of light add up to a light we can see, and we call it the aurora or the northern lights."

"Humm. What's an electron, and what was it you called those other things, atoms and molecules?"

"An electron is a tiny piece of matter that carries negative electrical charge. There are also protons, and these carry positive charge. If you put an electron and a proton together they become a hydrogen atom. When two hydrogen atoms join together they form a hydrogen molecule. Everything around us is like that, all made up of atoms and molecules, and these are built up of electrons and protons [and also neutrons, but no need to mention that now]. The air we breathe is mostly nitrogen and oxygen molecules, and up where the aurora is, the air is very thin and it is composed of oxygen atoms and molecules, nitrogen atoms and molecules and hydrogen atoms and molecules, plus some other things like that."

The father knows that his daughter probably did not comprehend all that he told her, but she at least she did not get any misinformation. He hopes that the conversation may have prepared her to fend off some wrong explanations she will later hear or read: that the aurora is a reflection of sunlight glinting off icebergs, off polar icecaps or off northern oceans. Explanations such as these were abandoned by scientists more than a century ago, but they have persisted in lore and literature up to the present time. Even in the 1980s, it was possible to read in a newspaper feature on natural phenomena syndicated by the Walt Disney organization that the aurora is the reflection of sunlight from ice crystals in the air. If Walt Disney were still alive, he would have been horrified by the feature because he always wanted things to be right. Had he chosen to portray the aurora, he might have directed a cartoon movie that showed electron bees buzzing down into the air from above to sting the noses of Santa's reindeer to make them glow green or cherry-red while that old gentlemen drove them along a wavering path across the sky, calling out, "Here Nitrogen, here Oxygen, with your glows so bright, light my way across the polar snows tonight."

As the man hinted to his daughter, the aurora is a light emitted (given off) from the gases in the earth's high atmosphere, mostly oxygen and nitrogen, when they are struck by fast-moving electrons and protons entering from above. By virtue of their motion, the electrons and protons carry kinetic energy (the energy of motion). This energy originates on the sun and is carried to the vicinity of the earth by electrons and protons. There, processes act to speed up the electrons and protons, and earth's bar magnet-like magnetic field guides them into the atmosphere near the polar regions.

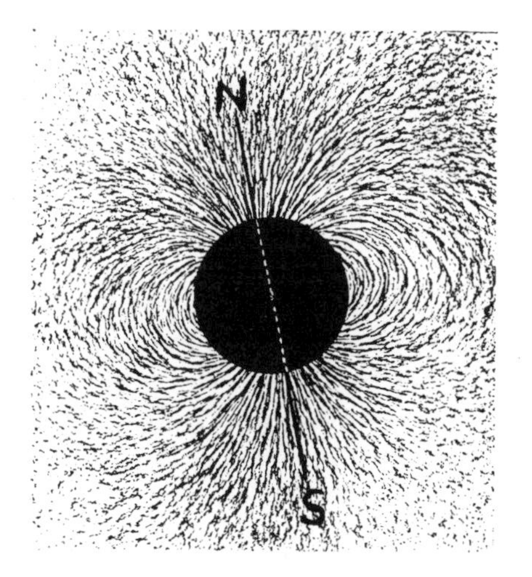

Figure 1.1 Iron filings near a magnetized sphere show the configuration of the earth's magnetic field—the same as that of a bar magnet.

An electron or proton coming down into the atmosphere typically strikes many atmospheric atoms and molecules before it loses all of its energy of motion and finally comes to rest. The collisions give the atmospheric atoms and molecules extra energy that they soon shed by emitting little flashes of light called photons. Photons are to a bright light as grains of sand are to a beach: both the light and the beach are made up of many little individual units, photons in the light, and grains of sand in the beach. Each light photon is of a specific color determined by the characteristics of the material giving off the photon. Oxygen atoms and nitrogen molecules produce most of the auroral light we see, so the nature of these atmospheric gases largely determines the colors of the aurora.

For a mental image of the auroral generation process, think of the incoming fast particles (the electrons and protons) as cue balls that strike other balls (the resident oxygen and nitrogen atoms and molecules) and then go on to strike yet others. Then imagine that instead of rolling across the pool table, the struck balls each give up their newly acquired energy by giving off a flash of light (a photon). However, each individual flash emitted in the aurora is too weak to be seen by the human eye. Many millions

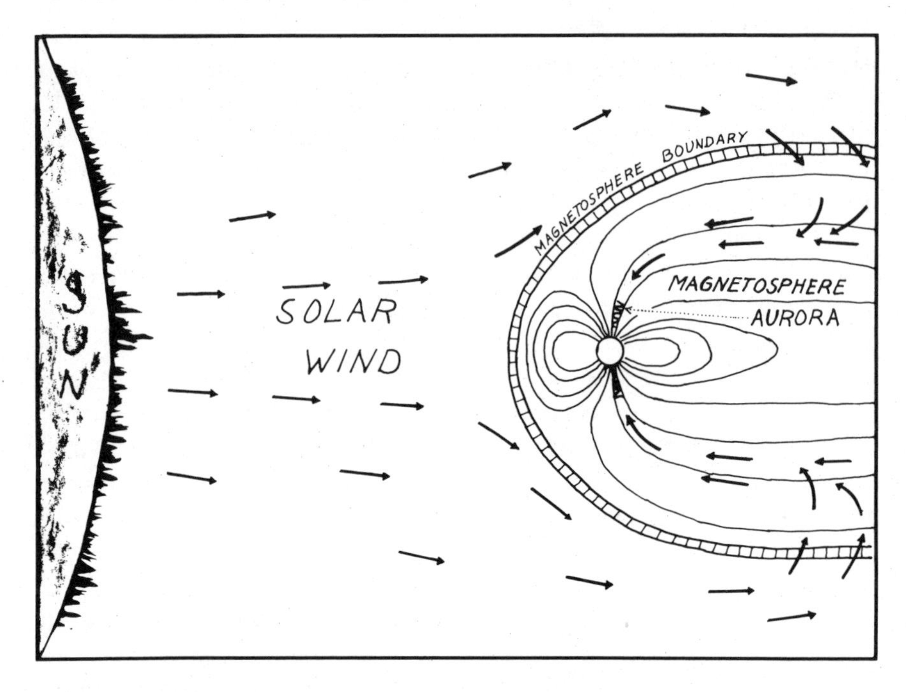

Figure 1.2 Charged particles flow out from the sun in the solar wind, skirt the boundary of the magnetosphere, and some enter well behind the earth where the magnetic field guides them down into the polar atmospheres.

of flashes must be emitted each second within each cubic centimeter of a large volume of the high atmosphere to yield a visible aurora. The incoming flow of fast particles must amount to at least 100 million particles per second over each square centimeter of sky area.

The aurora is an ever-present part of the earth's environment. It waxes and wanes, but it is always occurring somewhere, and it goes unseen in daytime because of obscuring sunlight. At certain times, during periods known as magnetic storms, huge flows of incoming electrically charged particles enter the atmosphere from the sun. Auroras then become widespread, and may, in rare cases, extend from the polar regions toward the equator far enough to fill the skies over as much as two-thirds of the earth's surface. Only during these comparatively rare events, known as great magnetic storms, do persons living in the warmer parts of the earth get to see the aurora.

An easy way to understand the general nature of the overall process that creates the aurora is to compare the sun-earth system with the television tube in an ordinary TV set. The back end of the television tube contains a cathode that boils off electrons. An electric field inside the tube accelerates the electrons, which then stream through the television tube under the guidance of carefully controlled magnetic and electric fields that steer the beam to appropriate locations on the face of the television tube. There, the electrons hit a phosphor coating material that emits light when so struck. The end result is the picture seen by viewers.

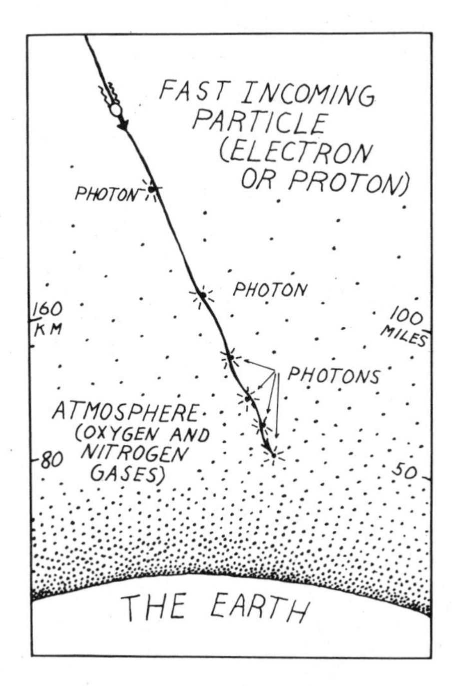

Figure 1.3 Fast incoming particles, electrons or protons, strike oxygen and nitrogen gases in the high atmosphere, causing them to emit auroral photons.

In this television tube comparison, illustrated in Figure 1.4, the portion of the system that corresponds to the electron-emitting cathode is the sun and its persistent outward extension, the solar wind. The sun continuously boils off electrons and protons (and far lesser amounts of other charged particles) which stream out away from the sun in the flow called the solar wind. The solar wind sometimes blows softly and at other times with gale force. As it approaches the earth, the solar wind runs into the earth's magnetic field which acts like a barrier. Thus, most of the particles flow past the barrier—it is called the magnetosphere—much as water moves past an obstructing rock in a river. However, the magnetosphere is not a perfect obstacle, and some of the charged particles in the solar wind seep across the outer boundary of the magnetosphere. Once inside the magnetosphere, the particles come under the control of the earth's magnetic field and any electric fields there. The electric fields speed up the charged particles, and the magnetic field guides them into the earth's polar atmosphere. The electric and magnetic fields determine precisely where the charged particles (the electrons and protons) will strike the atmosphere, just as do the fields inside a television that scan its beam onto the phosphor face of the tube. The atmosphere responds to the impact just as does the phosphor coating on the TV tube: it emits the light that an aurora watcher sees.

This analogy requires no stretching of the facts. The aurora is created in exactly this way, and a viewer of the aurora is, in effect, merely observing the face of a giant television tube that extends from the atmosphere to the sun's surface, 93 million miles (150 million km) away. However, most of the action that creates the intricacies seen in the aurora takes place near the earth, within the magnetosphere, an entity that extends out toward the sun a distance equal to approximately 20,000 miles (30,000 km). To tune into this show, a person near the auroral zone (the region where auroras occur most frequently) need only put on a parka and walk outside on a clear, dark night. The show may already be underway, or the person might have to watch for a while for it to begin.

In terms of concepts familiar to most people, the television analogy is perhaps the most meaningful way to describe how auroras occur. Another approach is to state that the aurora is an electrical discharge phenomenon. Two meanings of the word 'discharge' then apply. One relates to the light produced in the aurora: a glow discharge such as occurs in a neon tube. The term similarly applies to the light seen in a lightning stroke and the gap of a spark plug when air suddenly breaks down and a surge of current flows, creating both heat and light. The word 'discharge' also means to unload or to release. We speak of discharging a battery; the surge of current across the gap of a spark plug discharges a condenser, and a lightning stroke discharges a concentration of electrical charge built up in a storm cloud. Lightning strokes and sparks across a gap are brief events, but discharges can persist if a continuous source of electrical energy is available. The two wires that connect a neon bulb to a battery or an electrical generator maintain the voltage on the tube that allows it to glow continuously. The auroral glow also persists, and so it too must have a continuing energy source. The source is far away, and it has a formidable title. As the solar wind interacts with the boundary of the earth's magnetosphere, it forms a huge magnetohydrodynamic dynamo, a device akin to smaller and more simple types of electrical generators that supply electrical power. This distant generator develops a high voltage that feeds down into the auroral regions by means of the electrically charged particles that strike the atmosphere to create the aurora. No wires actually connect the gigantic electrical generator to the auroral atmosphere, but the earth's magnetic field forms a connection that is good enough so that the atmosphere acts as the main energy-dissipating load on the generator. Only a small part of the energy discharging into the auroral atmosphere actually generates auroral light; most of the remainder is lost to heating up atoms and molecules in the high atmosphere and to breaking them apart (ionizing them).

No need to think too much about this now. It's getting dark, and so it is time to go out and look at the aurora, photograph it, and perhaps even hear it.

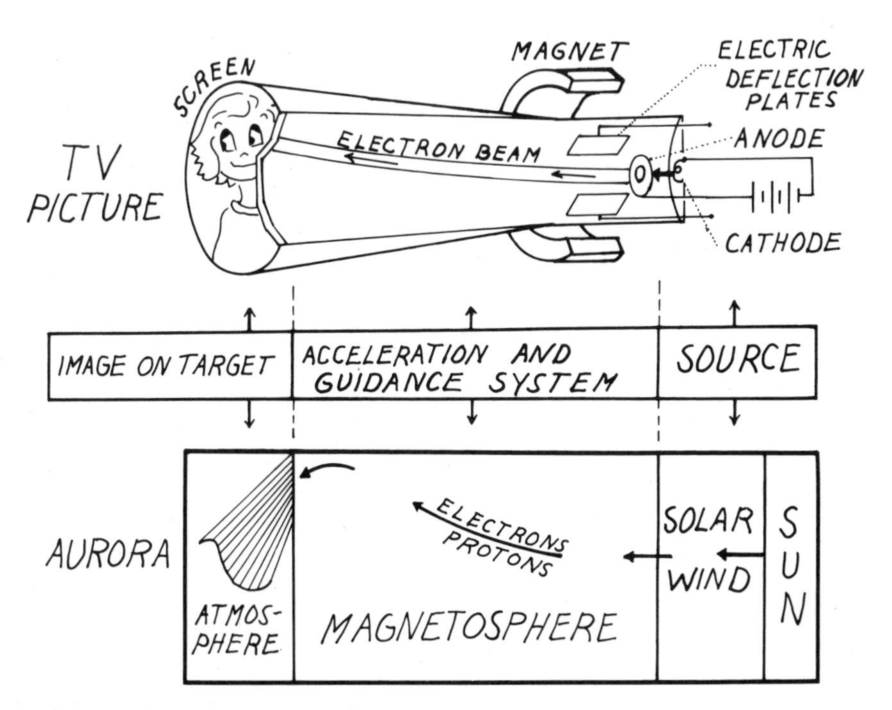

Figure 1.4 TV-magnetosphere analogy.

Section 2
HINTS FOR WATCHING AND PHOTOGRAPHING THE AURORA

Section 2.1 HINTS FOR WATCHING AURORA

The observing season in Alaska: From about August 15 to April 15. (Farther south, where darkness pertains year around, aurora is seen in summer also. At these locations, the aurora may occur slightly more often in spring and fall than in summer or winter.)

Best locations in Alaska: Any place north of the Alaska Range. The Fairbanks area is excellent, and Circle and Fort Yukon (fly in only) are better yet. Southern Alaska (for example, the Anchorage area) is only fair. A person's chances of seeing aurora are better in the northern tier of states in the Lower Forty-eight than in the Aleutians or perhaps even in Southeast Alaska where skies are often cloudy.

Best locations in Canada: Central Yukon Territory; the area near Great Slave Lake; northern parts of Saskatchewan, Manitoba and Ontario; central and northern Quebec; and Labrador. Other good observing locations in the northern hemisphere are at the south tip of Greenland, all of Iceland, and northern Scandinavia.

When to look: Any time from evening twilight until dawn.

Where to look: Usually in Alaska, aurora is first seen toward the northeast, but do not ignore the overhead sky, especially in early fall and late spring when sunlight obscures aurora close to the northern horizon. In eastern Canada, the first aurora seen is likely to be due north.

Best times to look: During any clear, dark night, but moonless periods are preferable. Statistically, the few hours near midnight are best. If no aurora is seen early in the evening, do not give up. It may appear an hour or two later—at any time of the night.

Best nights to observe:

1) The next night after someone has reported seeing a good aurora.

2) Nights 27 days after nights of major auroral displays (because, as seen from the earth, the sun appears to rotate on its axis with a period of 27 days, and its long-lasting active regions can repeatedly enhance auroras, every 27 days).

3) Nights during periods when high levels of magnetic activity are forecast. (For information, contact the Geophysical Institute in Fairbanks; telephone to (303) 497-3235 to receive a tape-recorded forecast prepared by the Space Environment Services Center in Boulder, Colorado; or, 18 minutes after the hour, listen to the forecast by tuning in the WWV time signals broadcast on 5, 10 and 15 MHz.)

Good observing locations: Any place with a clear view of the sky (especially to the northeast in Alaska), and where city or auto lights do not interfere.

Dark adaptation: Give yourself a chance by letting your eyes adapt to the dark. Every minute helps, but it takes 15 or 20 minutes for the eyes to become fully dark adapted after being exposed to any bright light. You might see a bright aurora right away, but weak ones might go unseen if your eyes are not dark adapted. If you need to go into a lighted room or have to look at a passing car's headlights, keep one eye closed to retain its dark adaptation.

Aurora or clouds? Even experienced observers sometimes have trouble distinguishing between clouds and weak auroras. The key is to look for rapid changes or internal motions. If they occur, it is aurora rather than cloud.

Colors to expect: Although they occur rarely—during times of exceptional magnetic activity—some auroras are blood-red. Most often, the observer in Alaska will see the 'normal' green aurora, but any very weak aurora appears colorless to the eye. As the brightness increases, the green color becomes apparent, and bright green auroras may have red lower borders.

Sometimes yellowish hues appear, and some people also recognize bluish colors.

Motions: A characteristic of auroras is that they move. Weak auroras typically move slowly, but bright ones can move remarkably fast. If you notice an increase in the degree of motion, keep watching during the next few minutes, because the aurora is likely to become more spectacular.

Aurora Watching from Airplanes and Autos: If you are travelling from the Lower Forty-eight to Alaska by air during the dark hours, try to obtain a window seat on the right-hand side of the aircraft (or if travelling southward, sit on the left). Sometimes the pilots will announce when aurora is visible, but not always. Eliminate interference from aircraft cabin lights by placing your face close to the window and blocking out the lights with a blanket or pillow. If you do see aurora, some of it may appear to be below the wing of the aircraft. This aurora actually is well above the altitude of the airplane, and is located far to the east over western Canada. Those who travel the Alaska Highway during fall, winter and early spring have an excellent chance of seeing aurora during the dark hours. It usually will be located on the right-hand side of a north-travelling vehicle. For the best views, pull over to the side of the road and turn off the headlights.

Section 2.2 HINTS FOR PHOTOGRAPHING AURORAS

1. Always use a tripod and a shutter-release cable because time exposures are necessary.

2. For black and white photographs use panchromatic films rated at ASA 100 or higher. Good color photographs can be taken with Kodak's Kodachrome ASA 200 film, Kodak Ektachrome ASA 200 or ASA 400 films, Fuji ASA 400 film or Konica 3200 film. Some color films of high ASA rating (for example Ektachrome ASA 800 and ASA 1600) may actually produce worse results than the slower films because of graininess and reciprocity failure (the tendency to become less sensitive with time during the exposure). Some films may be pushed to higher speeds with special development. It is best to consult a film processor about this, and also to remember that all exposures on a roll will receive the forced development. On a recent Shuttle flight the astronauts acquired excellent auroral photographs—ones that showed the green color of aurora especially well—with Kodak Ektapress ASA 1600 negative film.

Do not be surprised if color photographs fail to yield proper hues. These films are balanced for sunlight, and the auroral spectrum is quite different. Overexposure of color films creates white images.

3. Required exposures depend on film speed, lens opening and auroral brightness. The range is 1 to 20 seconds.

4. Focus on infinity and use the widest lens opening, unless you wish to bring more of the foreground into the picture. At f/3.5—the slowest lens setting you should use—the foreground will usually show up better than with a wider lens opeing.

5. The most pleasing photos are those with some landscape in the foreground, but avoid bright lights in the foreground, as they will tend to be overexposed in the long exposure required for the aurora. A slight amount of moonlight will help. Use of a wide-angle lens better allows including the foreground and helps bring it into focus, but remember that these lenses are slower than those with normal fields of view.

"Dress warmly…and take notes." (Carl Störmer at the camera.)

6. Bright and highly structured auroras tend to move fast, thereby blurring the image; the shorter the exposure the better, generally but a few seconds. For black and white photographs, a good rule is, "shoot until it moves."

7. The automatic exposure systems on some cameras may work reasonably well if the system is of high quality and if the camera remains warm enough to keep battery power high. However, these systems tend to overexpose auroral photographs, so manual setting of exposure usually gives better results.

8. The surest method of obtaining proper exposure is to shotgun the exposures; that is, take several exposures ranging from 1 to more than 10 seconds.

9. Below 0° F, keep the camera warm with a cover of some sort except while shooting. Do not move the camera repeatedly from outdoors to a warm room because water will condense on the cold lens when it is taken indoors. One way to avoid this problem is to place the camera in a zip-lock plastic bag before bringing it indoors, and then allowing the camera to warm up before removing it from the sealed bag.

"...keep the camera warm with a cover of some sort..."

10. Beware of static discharge which can damage film, especially when it is cold and dry. Wind films slowly and, if possible, ground the camera to earth or a large metal object. Try to avoid using film that has remained in the camera for days or weeks because it may have dried out and be more prone to static discharge.

11. Photographs taken when the moon is bright will have low contrast and a "washed out" appearance.

12. Try to anticipate so as to avoid changing film when the aurora is most spectacular. Changing film on a cold, dark night is not pleasant.

13. For future reference, keep notes on exposure times.

Home video cameras are not sensitive enough for auroral photography. Television is a powerful tool for studying aurora, but the technique requires the use of expensive, highly specialized cameras.

Remember also that the aurora is far away, and that it is an actual source of light. For both reasons, the use of flash or flood lights will not shorten the exposure time needed to capture the aurora on film or assist video recording.

Section 3

A SUGGESTED OBSERVING PROGRAM FOR THE AURORA WATCHER INTERESTED IN INVESTIGATING AURORAL SOUND

In this era where most new information is obtained by instrumental means, amateur aurora watchers are not likely to contribute to the general advance of knowledge—except that they may be able to help solve the mystery of auroral sound. A general discussion appears near the end of this handbook, in Section 17, and it should be easily comprehensible even without reading the intervening material. Here, in association with the hints for general aurora watching and photography, is a suggested program for making useful observations relevant to the mystery of auroral sound:

1. Keep a notebook, and record the timing of observations as accurately as possible. (Remember that routine data-taking, including auroral photography with all-sky cameras, may be underway by scientific organizations, so, if accurately timed, your observations can be keyed to the other information.) List all the times you are observing, noting auroral behavior and both the occurrence and absence of sounds, either identified or unidentified.

2. When first starting to observe, spend a few minutes listening for whatever sounds might be in the area, noting the results in your notebook. Also record the general observing conditions, paying particular attention to air movement, temperature and sky clarity.

3. Intensify your observations whenever aurora appears, perhaps every few minutes making a conscious effort to listen for sounds. Again, record the results in your observing log.

4. If the aurora brightens or if bright aurora appears to be moving toward the overhead sky, try to observe continuously. Now would be a good time to turn on a tape recorder if one is available; it might record auroral sound. If the recorder has a remote microphone pickup, place it as far from the machine as possible to minimize recorder noises.

5. When and if a potentially auroral sound is heard, it might last only for a few seconds. Then you will be able to do little more than determine the nature of the sound and the concurrent auroral behavior, and you should also note the exact time. If the excitement of the moment permits, try sniffing the air for any unusual odors. Determine if any persons with you also heard the sound and get their descriptions. Write everything down in your log as soon as possible.

6. Suppose that the sound persists. Close your eyes to determine if the sound is still heard. Cover your ears to see if it persists. Examine the auroral behavior carefully in an attempt to determine the degree of correlation between the aurora and the sound's levels and characteristics. Try to determine if the sound appears to come from any particular direction. Move around, perhaps putting an ear close to the ground or other objects to see if the sound persists and where it might be coming

Turn on your tape recorder, sniff the air, watch the aurora, take notes, acquire photographs.

from. Continue to sniff for odors, and examine the surroundings for any evidence of nearby light such as Saint Elmo's fire.

7. Whether or not you heard auroral noises during a program of careful observation, the results written in your log may be of value to an organization such as the Geophysical Institute of the University of Alaska Fairbanks. Copies of the appropriate pages of your log, extracts of them, or a brief statement of the overall results of your observations will be appreciated, and if you managed to record sound on tape, the tape is extremely valuable, and copies should be made. Since auroral sound is yet to be recorded, you might have a first, and analysis of the tape by a scientific organization will be in order.

Section 4
SOME BASIC FACTS AND DEFINITIONS

Section 4.1
THE AURORAL ZONE

In each hemisphere, auroras occur most frequently in an elongated belt called the *auroral zone* that encircles the polar region (see Figure 4.1). The *peak of the auroral zone* (Figure 4.2) is a statistically determined imaginary line drawn across the earth's surface. If an observer stands on that line for a long time, several years, he or she will see more aurora than if anywhere to the north or south. Notice in Figures 4.1 and 4.2 that the peak of the auroral zone intersects the west coast of Alaska near Point Hope. It dips southerly as it crosses Alaska, passing almost directly through Bettles and Fort Yukon and on across the Canadian border to the north of Dawson. The peak of the auroral zone passes near Yellowknife then, to the east, crosses the juncture between Hudson Bay and James Bay. It enters the Labrador Sea to the north of Goose Bay. As it circuits the globe, the northern auroral zone forms an approximate circle 23 degrees of latitude in radius (2500 km or 1600 miles) centered on the geomagnetic pole (defined elsewhere) near Thule, Greenland. Notice that as the auroral zone crosses North America, from west to east, it trends southerly. For that reason, residents of the northeastern states see more aurora than do those living more westerly along the United States-Canada border. To see aurora frequently, an observer need not stand precisely on the line defining the peak of the auroral zone, because a person

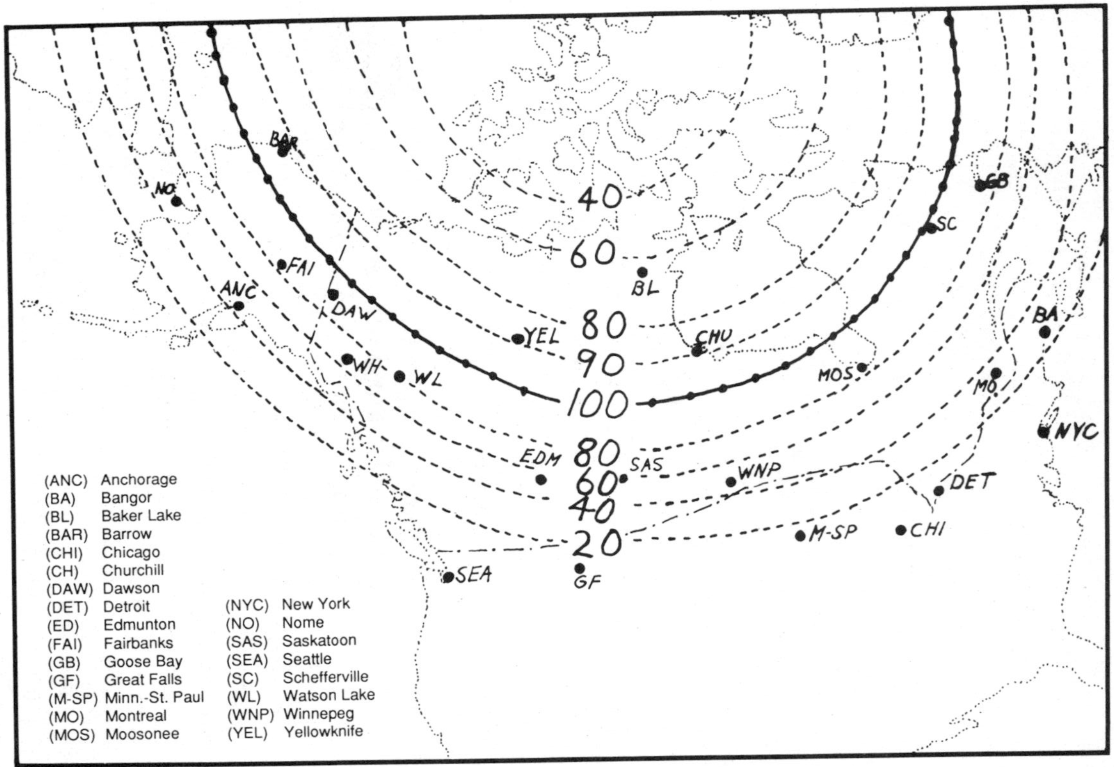

Figure 4.1 The long-term percentage of dark, clear nights on which aurora can be seen at locations in North America. The peak of the auroral zone (see also Figure 4.2) is the heavy dotted line labled 100.

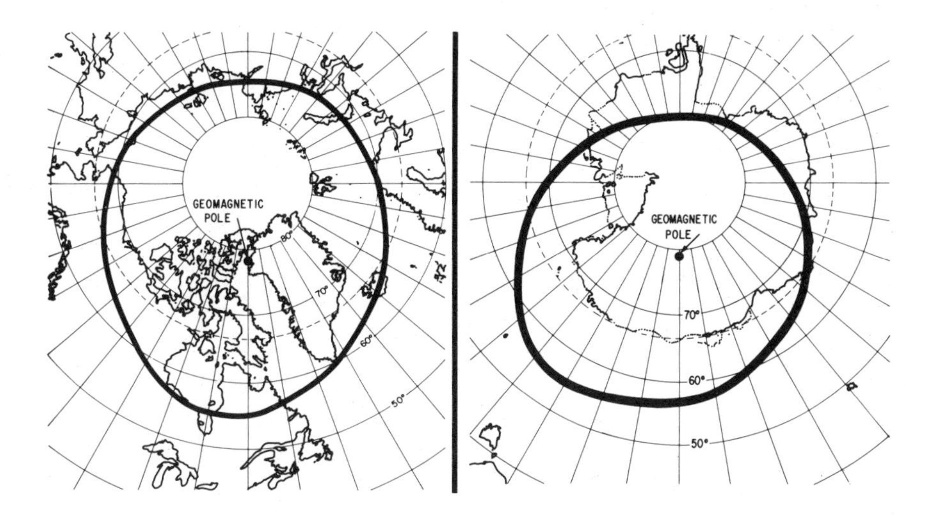

Figure 4.2 The peaks of the two auroral zones, centered about the northern geomagnetic pole near Thule, Greenland, and the southern one at Vostok, Antarctica.

located 100 or 200 km to the north or south will see almost as many. Thus, though located approximately 200 km south of the peak, Fairbanks is an excellent observing location.

Section 4.2 THE AURORAL OVAL

The auroral zone is a statistical concept, whereas the auroral oval is an entity that can be observed directly. The *auroral oval* is the region containing essentially all of the aurora observed at any instant in the northern or southern hemisphere. (The word 'essentially' appears here because some auroras occasionally occur in the polar cap region located inside the oval, sometimes seemingly disconnected from those within the oval.) An earth-bound observer can see only a fraction of the auroral oval because it extends in quasi-circular fashion around the entire polar region. The oval behaves as though it were attached to the earth at the geomagnetic pole, but the oval is skewed toward the nightside of the earth, as shown in Figure 4.3. During times of low auroral activity, the oval is narrow and somewhat contracted in toward the geomagnetic pole. As the level of auroral activity increases, the auroral oval widens and its diameter increases.

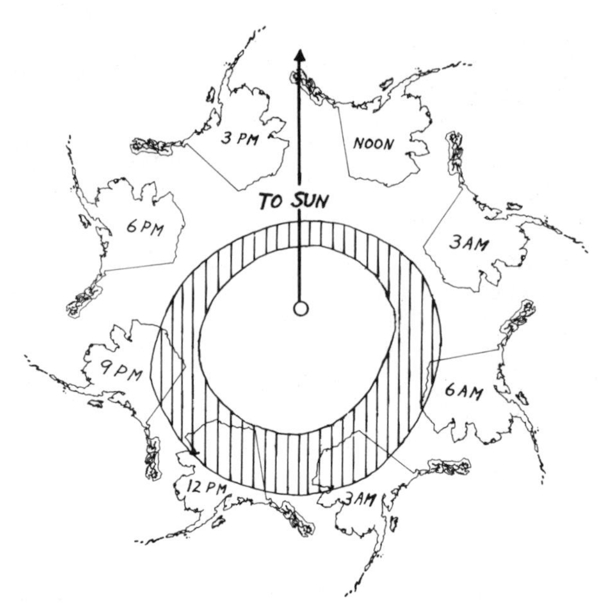

Figure 4.3 A statistical rendition of the northern auroral oval. As the earth rotates beneath, the oval remains in a fixed orientation relative to the direction toward the sun (arrow), and has off-center "attachment" to the earth at the geomagnetic pole (encircled dot at the base of the arrow). Here, the changing positions of Alaska relative to the fixed oval demonstrate how the earth's rotation swings this and other parts of North America beneath the oval only during the night hours.

Section 4.3 AURORAL DISPLAYS

The term *auroral display* is somewhat ambiguous in that it can have several meanings. It usually refers to the collection of auroras seen at one location at one time. "I saw a nice auroral display at 9 o'clock last night," someone might say. Or he might say, "We had an extensive auroral display last night. It lasted from dusk until dawn." Here, the speaker is referring to the aurora seen over a span of time. The distinction is not important, since the context of the speaker's words will convey what he or she means.

Section 4.4 AURORAL FORMS

Auroral forms are the individual curtain-like structures that make up an auroral display. Years ago, when most scientific observation of aurora was performed visually, scientists developed a nomenclature to describe auroral forms. Since this was the pre-satellite era, the scheme naturally hinged around the appearance of the auroral structures as seen by a person on the ground. Were one to devise a new scheme today, it might be much different. The original scheme describes the apparent shape of an auroral structure, any internal regularities it might contain, the structure's brightness and color, and in some cases even its location in the observer's sky. Auroral arcs and bands are the most common forms seen by most observers—an arc is a form with uniform curvature, and a band is merely a kinky arc, one that can take on a variety of shapes. Arcs and bands may appear relatively homogeneous, or they may contain prominent near-vertical striations called rays.

Section 4.5
HOW TO LOCATE AURORAS AND OTHER PHENOMENA IN THE SKY

An observer looking up into the sky has no immediate way to judge distance. Many faulty reports on sky phenomena come about because people do not recognize this fundamental limitation. Quite commonly, people will report bright meteors as landing "just beyond the trees," or "just over the hill," when in fact they did not actually see the meteor land. It merely passed over the horizon while still many miles above the earth's surface. The same is true of auroras, which rarely if ever come closer to the earth than about 60 km. Where an aurora appears to touch the horizon it is about 1000 km away, and typically at altitude near 100 km. Yet some people believe that they have seen aurora within nearby trees or between them and a distant hill. The mistakes are understandable because a combination of sky clarity and the crispness of the outlines of some auroras sometimes makes it seem impossible that the forms are more than a short distance overhead. "It looked as though I could reach out and touch it," is not an uncommon remark.

Since the determination of distance to faraway objects is impossible for visual observers, the only useful way to describe the location of an aurora or another object in the sky is by citing its angular position. "It was 25 degrees above the northeast horizon," a person might say. He could also have said, "It was to the northeast at zenith angle 65 degrees." In the latter statement the observer is using the *geographic* (or *true*) *zenith*—the point in the sky vertically overhead—as the starting point for his measurement. The two

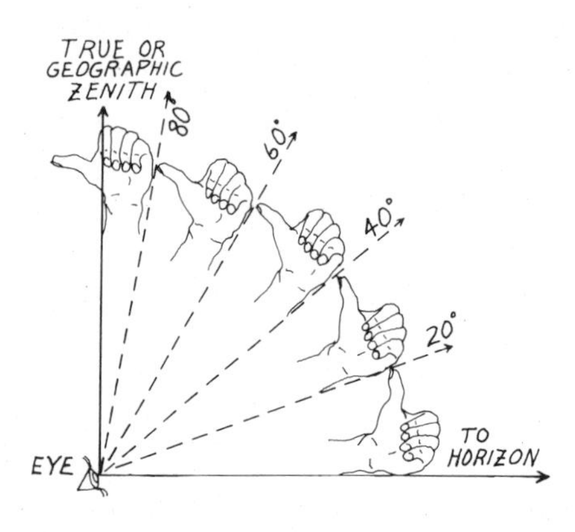

Figure 4.4 Angular measurement in the sky using the fully outstretched arm and thumbs-up clenched hand.

statements are equivalent, but if an object is near the horizon a measurement of its elevation angle (angle above the horizon) is easier than measurement of its zenith angle.

An always available tool for estimating angular distance in the sky is the human arm and hand. If, as is illustrated in Figure 4.4, a person stretches out an arm and places the bottom of the clenched fist on the horizon, the end of the thumb is 20 degrees above the horizon. By "walking" the hand up the sky, a person can find the approximate elevation angle of any object, and also verify that 4-1/2 "steps" brings the hand to the geographic zenith.

Other useful indicators of angular distance in the sky:

Diameter of the moon: 1/2 degree.

Diameter of a 25-cent piece held at arm's length: about 2 degrees.

Distance between the two Pointers of the Big Dipper: 5 degrees.

Distance across the open side of the cup of the Big Dipper: 10 degrees.

Distance from the Pole Star to the nearest of the two Pointers: 28 degrees.

The horizon always provides a useful plane from which to measure angles in the sky. An auroral watcher will soon recognize another important starting point that occurs high overhead, the *auroral coronal point*. The auroral coronal point is a point in the sky toward which the rays contained within some auroral forms appear to converge. It is located exactly in the magnetic zenith, the point in the sky the observer views when looking along the direction of the earth's local magnetic field. Thus, 'magnetic zenith' and 'auroral coronal point' are two names for the same point in the sky, hereinaf-

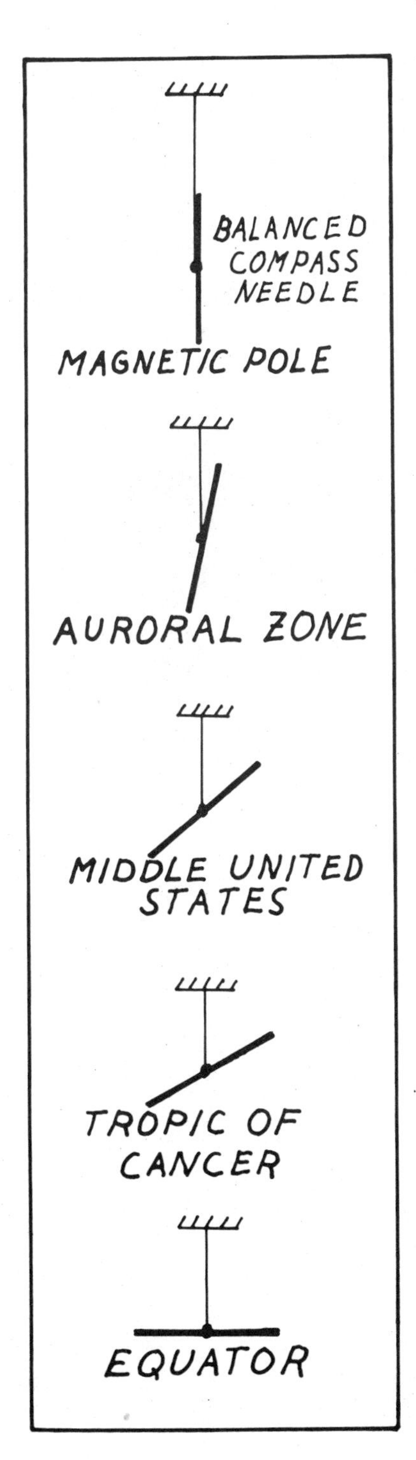

ter usually referred to as the magnetic zenith.

If a thread is fastened exactly to the middle of a magnetized needle and the needle suspended on the thread, the needle will align itself in the direction of the earth's magnetic field. At the earth's magnetic equator the needle will hang horizontally, but if moved away from the equator, the needle will dip downward on one end. The other end points toward the magnetic zenith.

If the needle is taken to the north magnetic pole, located in northern Canada, it will hang exactly vertical. Only here (and at the south magnetic pole) does the magnetic zenith coincide with the geographic (true) zenith. Except in the region between the north magnetic pole and the north geographic pole, the magnetic zenith in the northern hemisphere is south of the geographic zenith by an angle that depends on the observer's magnetic latitude, as is shown in Figure 4.5. The angle between the true and magnetic zeniths varies with magnetic latitude as follows: Magnetic

Figure 4.5 (at left) A magnetized needle hung by its balance point seeks out the direction of the local magnetic field, and so points to the magnetic zenith.

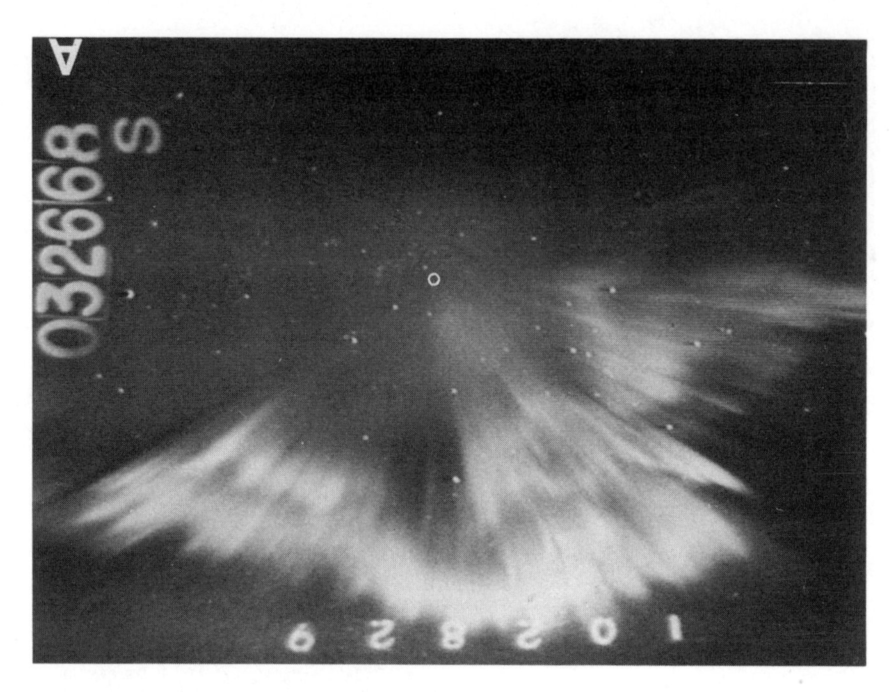

Figure 4.6 A television photograph showing how auroral rays point to the magnetic zenith (the white circle drawn in).

latitude 90 degrees (the magnetic pole), 0 degrees; at magnetic latitude 80°, 5 degrees; at 70°, 12 degrees; at 60°, 20 degrees; at 50°, 28 degrees; at 40°, 37 degrees; at 30°, 45 degrees; at 20°, 60 degrees; at 10°, 75 degrees; and at 0° (the magnetic equator), 90 degrees.

At Fairbanks, Alaska, the magnetic zenith is located approximately 13 angular degrees southwest (magnetically south) of the geographic zenith. The magnetic zenith at other Alaska locations ranges from 12 to 20 degrees southwesterly of the geographic zenith. Unseen in a sky without aurora, the location of the magnetic zenith becomes obvious when auroras that contain striations appear overhead, as in Figure 4.6. These striations are mutually parallel and very long, so they appear to converge to a point (in this case conveniently called the coronal point, but remember, it and the magnetic zenith are the same), just as do the rails on a long straight stretch of train track, or the parallel walls of tall buildings as seen from a street below (Figure 4.7).

The magnetic zenith is important to aurora watchers because the apparent shapes of most of the auroras they see are critically dependent on the auroras' locations relative to the magnetic zenith. An auroral form that crosses through an observer's magnetic zenith usually looks quite different than

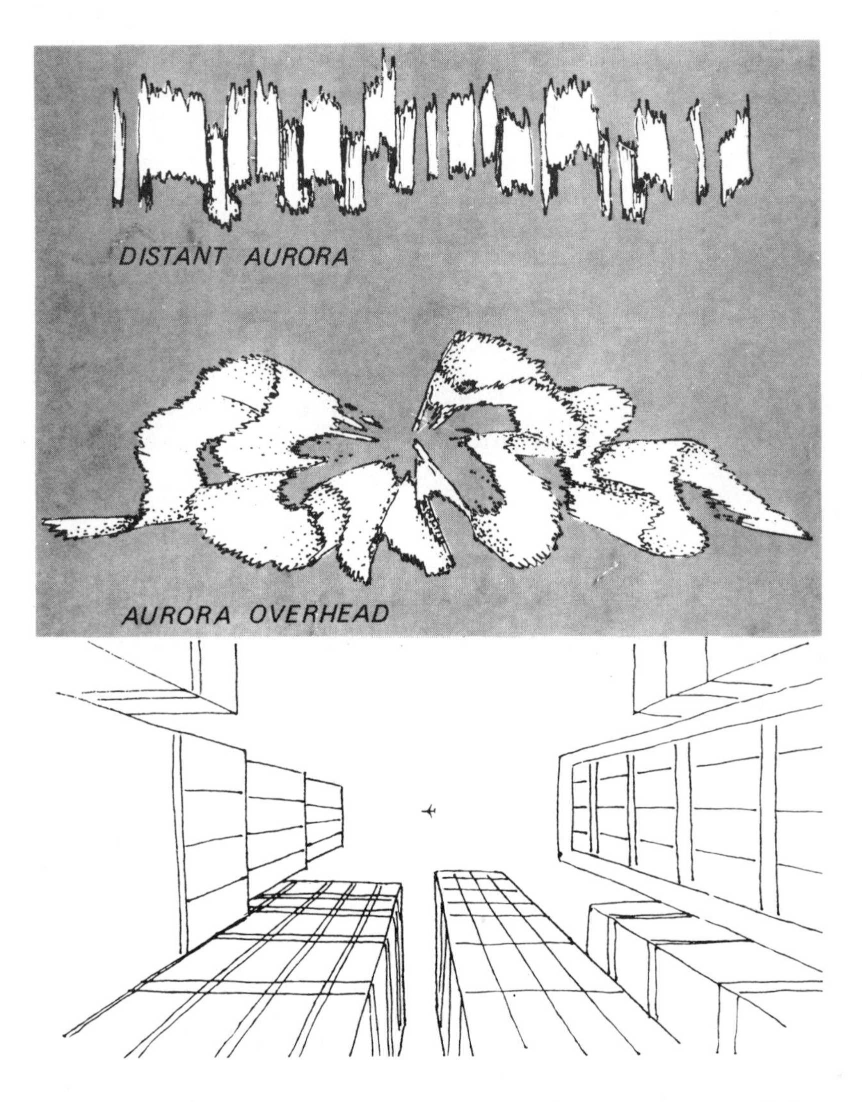

Figure 4.7 Like the walls of tall buildings when seen from the street below, the rays of an aurora appear to converge to the point overhead, while those of an aurora low on the horizon appear more parallel. (Drawing by Russell Mitchell.)

the same form seen only a few angular degrees removed. And like balanced compass needles, auroral rays align themselves precisely along the direction of the local magnetic field. Thus, in principle, if an observer sees a ray in his magnetic zenith it should look like a point or spot, but if the ray moves only a short distance from the magnetic zenith, it appears to be a line.

Nearly all well-defined auroral forms are extended in height and aligned along the direction of the magnetic field. The only practical way to describe their location in the sky

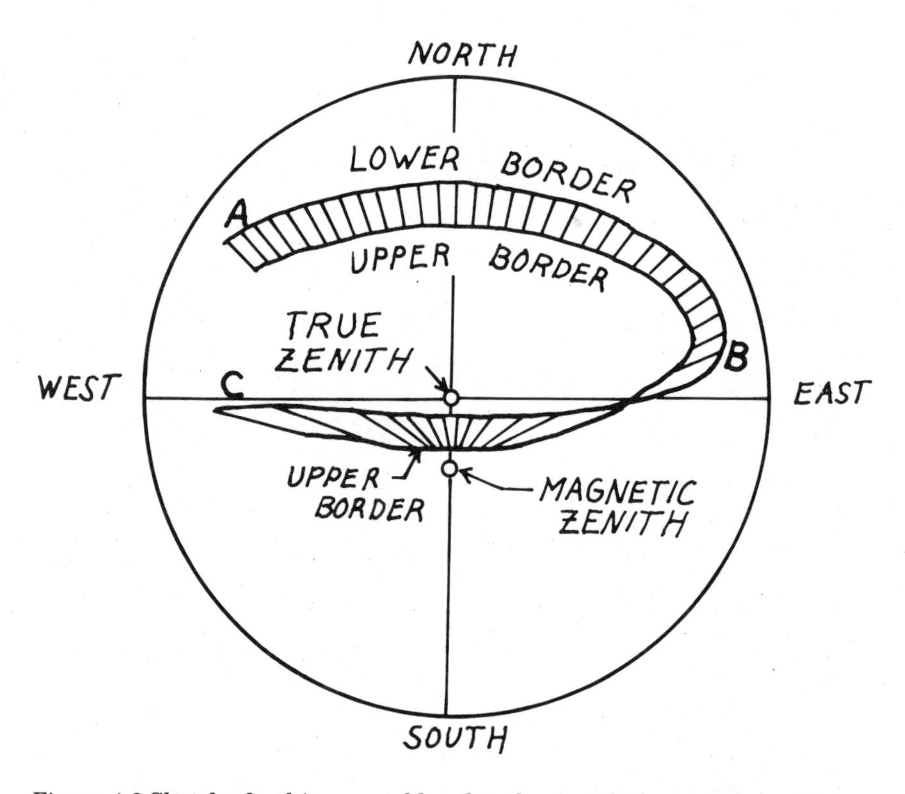

Figure 4.8 Sketch of a thin auroral band such as might be recorded with an all-sky camera located below it. The outer circle represents the horizon; the true zenith is at center, and the magnetic zenith is to the south of it. The band's lower border is the line ABC, the edge of the aurora that lies farthest from the magnetic zenith. Notice that in this case the actual thickness of the band can be seen only where it lies due magnetic east of the observer. See also Figure 4.9.

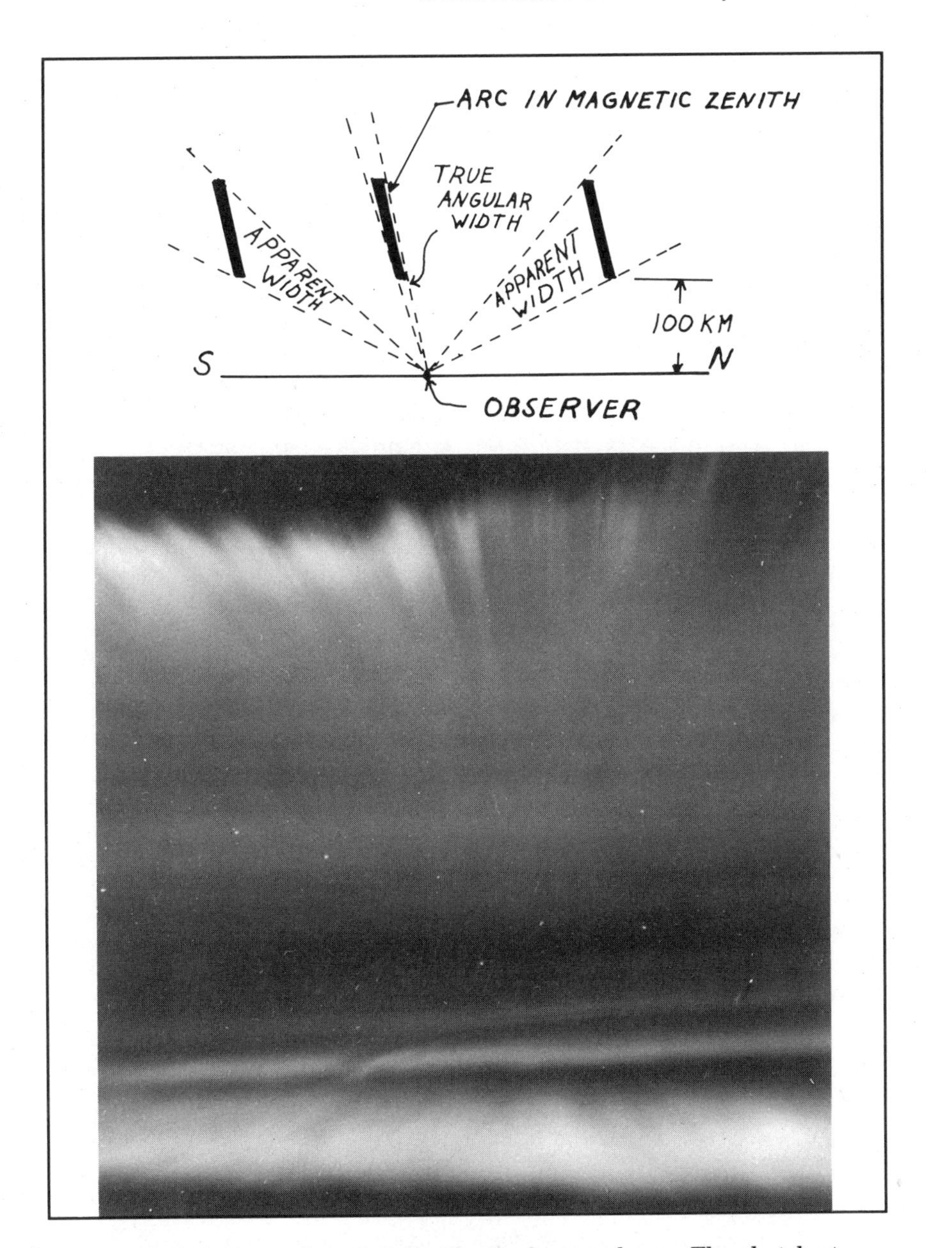

Figure 4.9 Two views of essentially identical auroral arcs. The sketch at top is a view from far to the east, showing how the actual width of an arc can be determined only when it is exactly in the magnetic zenith. The photograph at the bottom shows parallel arcs straddling the magnetic zenith, the third up from the bottom only displaying its actual width. Note the Big Dipper at center. (V. P. Hessler photograph.)

is to specify the angular position of their *lower borders* (their bottom edges). Because the altitude of the lower border is known approximately, it then becomes possible to estimate how far away the auroral form is and to plot its position on a map. An observer usually can determine which edge of an auroral form is the lower border because that edge is brighter and more distinct than the upper border. Sometimes, however, the distribution of brightness within the form is such that the observer does not easily recognize which boundary of the form represents the lower border. The observer must then resort to a rule: *The lower border of an auroral form always is that apparent edge that lies farthest away from the magnetic zenith.* Thus if an imaginary line is drawn from the magnetic zenith to the auroral form, that line will first intersect the form's upper border, and the line will exit the form at its lower border (see Figure 4.8).

Figure 4.9 will help the reader to understand why so much stress is being placed here on the importance of the magnetic zenith to the aurora watcher. The photograph shows the near-zenith portions of an array of auroral arcs that are nearly identical in shape—very thin, yet greatly extended in height up along the direction of the local magnetic field. The magnetic zenith lies just below the cup of the Big Dipper, within the third aurora up from the bottom. Looking closely, a person may recognize that this form actually is a pair of two closely spaced arcs. The next aurora below it appears thicker, and the one below that thicker still. Similarly, the auroras in and above the Big Dipper have an 'apparent' thickness proportional to their angular distance away from the magnetic zenith. The accompanying sketch illustrates that the true width of an auroral form is impossible to determine unless the form is exactly in the magnetic zenith. Again, each of the auroral forms in Figure 4.9 actually has approximately the same thickness.

For the purpose of mapping the locations of auroral forms, it is sufficient to assume that the altitude of the lower border of the green aurora usually seen near the auroral zone is 100 km, and that the altitude of the lower border of the more rarely seen all-red aurora is approximately 250 km. Based on those assumptions, the horizontal distance in kilometers between an observer and the observed aurora he or she sees is:

Auroral Type	Measured Elevation Angle to Lower Border 0	10	20	30	45	60	70	80	90
	Distance in Kilometers								
Green	1100	450	250	170	100	60	40	20	0
All-Red	1800	1000	600	400	250	120	80	40	0

The table shows, for example, that the place where a typical green aurora appears to meet the horizon is 1100 km distant from the observer, but if its lower border is 45 degrees above the horizon, the form is a map distance of only 100 km away. A green aurora seen at Fairbanks with lower border 20 degrees above the northeast horizon is approximately over Fort Yukon. One with its lower border 5 degrees above the northeast horizon of Anchorage is approximately over Fairbanks.

Section 4.6 BRIGHTNESS OF AURORAS

Many years ago, scientific observers established a scheme called the International Brightness Coefficient (IBC) to describe the brightness of auroras. In itself, the scheme may be of little interest to the casual aurora watcher, but it does suggest the range of brightness that the observer can expect—from barely visible to a brightness that is much less than that of a bright fireball meteor or the moon's disk.

IBC I Aurora that has brightness equal to that of the Milky Way.

IBC II Aurora with the brightness of thin moonlit cirrus clouds. (About 10 times brighter than IBC I.)

IBC III Aurora with the brightness of moonlit cumulus clouds. (About 100 times brighter than IBC I.)

IBC IV Aurora that provides a total illumination on the ground equivalent to full moonlight. (About 100 to 1000 times brighter than IBC I.)

Note that the scheme is not quite internally consistent since the first three steps refer to the brightness of objects in the sky, but the fourth relates to the illumination created on the ground and therefore depends both on how bright the aurora is and how extensive.

Section 4.7
AURORAL COLOR PATTERNS

The human eye contains light sensors called rods and cones. The rods are more sensitive to light than the cones, but they do not distinguish color. Also, away from its central axis, the eye contains more rods than cones. For that reason, the very weakest lights detected are those off to the side, those that are said to be seen with 'averted vision.' Such weak light appears colorless. Not until enough light comes into the eye to trigger the cones is a person able to distinguish the color of the light. Thus, the weakest auroras seen always appear colorless, and they are most easily seen when the head is turned away so that the auroral light comes into the side of the eye. That is why a person sometimes recognizes a light off to the right or left but is unable to see it after turning the head to view the source more directly.

Even when enough light comes into the eye to trigger the cones, the recognized color of the light depends on the eye's sensitivity in the different parts of the visual light spectrum. Individuals differ in their ability to detect light of different colors, but the average human eye is most sensitive to yellow-green light, and only about one-fifth as sensitive to red or blue light. Partly for this reason, and partly because the aurora contains much green and considerable red light, most auroras are recognized as having these colors. (See Plate 1, to be discussed later in Section 7.)

The apparent color of an aurora can also depend on how far it is above the observer's horizon. Light from the aurora has to travel a long distance to reach the eye. The length of the path for an aurora low on the horizon is about 1000 km, and even for one overhead the path is at least 60 km. As light traverses through the atmosphere the air molecules scatter the light, some colors of light more than others. The air more severely scatters light in the blue end of the spectrum than in the red end, and of course the light that gets scattered back does not reach the eye. Therefore, distant auroras, those low on the horizon, are more likely to have a yellowish and reddish cast than those nearby, the auroras overhead.

The main patterns of auroral coloration:

Green auroras: the most common type seen near the auroral zone (Color Type C). (See Plates 2 through 9.)

Green auroras shading toward red at their tops: recognized less often (Color Type A). (See Plates 10 through 15.)

Green auroras with red lower borders: usually occur during the very brightest portions of displays (Color Types B and E). (See Plates 16 and 17.)

All-red auroras: usually occur only during very large displays when the aurora reaches to the lower latitudes (Color Type D). (See Plates 18 and 19.)

Blue and purplish auroras: occur when tall auroras are exposed to sunlight or bright moonlight (Color Type F).

See Section 7 for an explanation of the causes of the various auroral colors and an elaboration of the listed 'Color Types.'

Section 4.8
THE AURORAL BREAKUP; PROGRESSIONS OF AURORAL FORM TYPES DURING AN AURORAL DISPLAY

The auroral breakup is the most spectacular part of a moderate to large auroral display, the portion that every aurora watcher hopes to see. Breakups involve a brightening of auroral forms and a rapid change from homogeneous forms to rayed ones, and change of arc-like forms to those that are highly contorted and typically swirling across the sky, as is shown by the example of a breakup sequence in Figure 4.10. One or more breakups might occur in one night, but a moderate to small display is likely to have only one. The most probable time for it to occur in the aurora over Alaska is during the hours 10:00 pm until 3:00 am. (A breakup observed at one location is a part of the global auroral substorm variation described in Section 6.7; specifically, the breakup occurs as part of the expansive phase of a substorm.)

If aurora watchers located near the auroral zone—at Fairbanks, for example—see

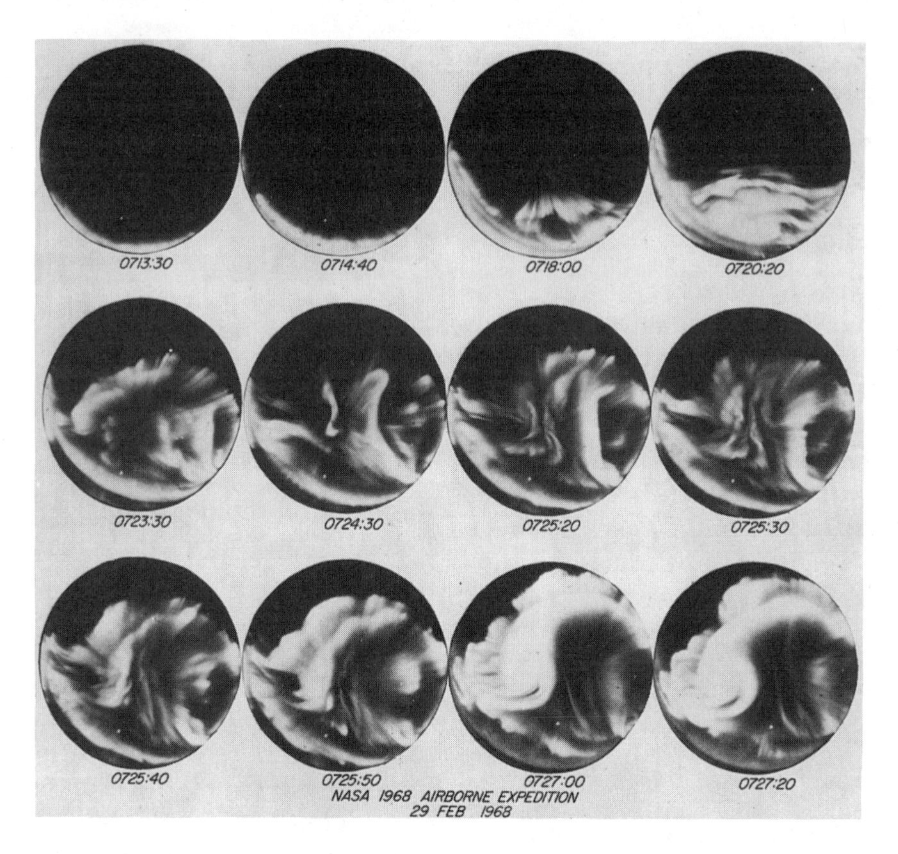

Figure 4.10 A series of all-sky photographs showing a break-up sequence. North is toward the top, east toward the right. Numbers indicate the times in hours, minutes, and seconds. The sequence spans about 14 minutes. (Courtesy of S.-I. Akasofu.)

multiple arc-like auroras during the evening hours they should keep a close eye on the aurora because, sooner or later, a breakup almost certainly will occur. If these multiple forms appear early in the evening, the breakup probably will be spectacular, and continuing observation through the night may reward the watcher with the sight of several breakups. Signs of impending breakup are a general brightening of the aurora, change from homogeneous to rayed forms, and an increase in the movements of rays along the forms.

Breakups observed during the early evening hours most likely will begin near the eastern horizon. Brightening of the aurora there typically moves westward, accompanied by swirling motions as arcs and bands become distorted and perhaps change to spiral-like configurations.

If a breakup occurs near midnight, the observer is likely to see a brightening in the southernmost portion of the aurora that progresses rapidly, sometimes in a minute or two, and a fast northward sweeping of the auroral forms. The sky may become covered with auroral forms that swirl and race about. Auroral forms tend to be brightest during the breakup, and they often have pronounced red lower borders. The fast motions, the brightness and the multiple colors produce a scene of great beauty.

Within a few minutes the breakup will be over, and the sky takes on a seemingly uninteresting appearance. However, the observer who continues to watch should soon see pulsating auroras, weak by comparison with the preceding aurora, but nevertheless fascinating to watch. The pulsations are likely to continue, perhaps for the rest of the night. If, while watching the pulsating aurora, the observer notices a tendency for the display again to form homogeneous arcs, she should continue to observe, for it is likely that another breakup is on the way. During highly disturbed periods (that is, during magnetic storms), incredibly beautiful auroras are sometimes seen very late in the night.

Section 4.9 THE INCOMING PARTICLES THAT CAUSE AURORAS

The electrons and protons that boil off the sun and eventually find their ways into the earth's polar atmosphere to generate the aurora begin their trip casually. Emitted sunlight takes only 8 minutes to arrive at the earth, whereas these particles drift outward in the solar wind so slowly (1000 times slower than the sunlight) that the journey takes several days. Once the charged particles enter the magnetosphere, they undergo acceleration to speeds near one-fifth that of the speed of light. They then are capable of penetrating into the atmosphere to a depth of approximately 100 km above the earth's surface.

Rather than talk of speed directly, scientists find it more useful to discuss the kinetic energy of the particles. The kinetic energy of a particle is proportional to the square of its

The Electron Volt, a Unit of Energy

If an electron were placed at rest in an electric field, the electron would begin to move along the direction of the field. Thus it would gain kinetic energy, the energy of motion. If the field were a weak one, say 1 volt per meter, by the time the electron moved 1 meter it would have passed along a voltage change of 1 volt, but it would have acquired a remarkable speed, about 600,000 m/s.

Electrons have very small mass, so a gravity field such as the earth's has little effect. And a magnetic field gives an electron or other electrically charged particle no acceleration at all. Only electric fields have a profound effect on the speed, and hence the kinetic energy, of electrically charged particles. Since only electric voltages radically affect the speeds of charged particles, a natural unit of energy (and speed) of a charged particle is the electron volt (eV). One electron volt is the amount of energy acquired by an electron moved through a voltage change of 1 volt.

A simple formula allows calculation of that electron's speed, and it is indeed a fast 600,000 m/s. However, the speed of light is 300 million m/s, so an electron with an energy of 1 electron-volt has a speed only about 1/500th the speed of light. An electron with energy 100 eV has a speed 1/50th the speed of light. An electron with energy 1 keV (1000 eV) has a speed of about 1/17th that of light, and a 10-keV electron moves at about 1/5th the speed of light. Protons with equivalent energies measured in electron-volts move much slower because they are 1836 times heavier than electrons, and therefore are harder for an electric field to accelerate to high speed. (The earth's gravity field, however, works much better on protons than on electrons because the protons are so much heavier.)

speed. As they arrive at the top of the atmosphere, the incoming electrons have kinetic energies in the range of about one to a few tens of kilo electron volts (keV), and the protons are in the range 1 keV to 100 KeV. (See box.)

An incoming auroral electron moving fast enough that it can penetrate the atmosphere down to altitude 100 km carries an initial energy of 10 keV. Called a *primary* electron, this particle will strike several hundred atoms and molecules in the upper atmosphere before finally coming to rest. During each impact the primary electron loses several to many electron volts, on average approximately 100 electron volts. Some of these impacts break up the target atoms and molecules, ejecting from each another electron. This produced electron, called a *secondary* electron, usually receives a portion of the primary electron's kinetic energy, typically about 90 electron volts. Initially essentially at rest, the secondary electron then also is likely to move rapidly down into the atmosphere, and before coming to rest, strike yet other atoms and molecules. Thus, the primary electron stream generates a cascade of downward-moving electrons composed of both primaries and secondaries. The secondary electrons, like the primaries, carry electrical charge and therefore travel under the guidance of the earth's magnetic field in identical fashion. (A detailed discussion of how a magnetic field guides electrically charged particles appears in Section 9. For now, it is sufficient to accept the fact that it happens.)

An incoming proton behaves somewhat differently than an incoming electron because the proton may grab an electron during its transit through the atmosphere. A proton attached to an electron is a hydrogen atom, an uncharged entity that does not come under the control of the earth's magnetic field. It then may move along in a random direction, and as it moves the hydrogen atom may have another impact that causes it to lose its electron, again becoming a proton. An incoming primary proton might go through several cycles like this, spending part of the time as simply a proton and another part of its time as a component of a hydrogen atom. The consequence is that auroral light produced by protons tends to be broader and less well defined than that produced by electrons. An auroral form seen to have sharp boundaries derives that sharpness from the incoming electron stream, not from the protons. The bright, highly structured auroras that an observer sees trace out the locations where copious flows of primary and secondary electrons are penetrating the atmosphere. They may be immersed in a faint, even unseen, emission of light caused by the incoming protons.

Section 5
KINDS OF AURORA

Section 5.1
MAIN TYPES: DISCRETE AURORA, PULSATING AURORA AND THE HYDROGEN ARC

Despite the many names applied to auroral structures—arc, band, drapery, corona, veil, diffuse surface, flaming aurora, pulsating surface, and several others—almost all observed structures really are only of three general types. One type is *discrete aurora*; as the name implies, discrete auroral forms are those that have sharp boundaries and are most easily seen. A second important category is *pulsating aurora* (also sometimes called diffuse and pulsating aurora), which exhibits quasi-periodic (0.1 sec to more than 20 sec) variations in brightness, but is always dim, and usually diffuse in appearance. A third type, quite distinct from the other two, is the *hydrogen arc* (or proton arc, also called the diffuse aurora), seen during the evening hours when observing moderate to large auroral displays. These three types of aurora have distinguishing characteristics that relate back to fundamental processes that create these auroras. (Before the distinction between pulsating aurora and the hydrogen arc was recognized, the two kinds of aurora typically were lumped together and described by the term 'diffuse aurora.' Later on, the term 'diffuse aurora' became specifically associated with the hydrogen arc.)

Section 5.2
DISCRETE AURORAL FORMS

Discrete auroras are the ones most often seen by casual observers, mainly because they are the brightest and most sharply defined, and they may be spectacularly colorful. They also make up the bulk of an auroral display seen during the premidnight hours, when casual observers are most inclined to look at the sky. One of the most important things to realize is that all discrete auroras share one fundamental characteristic: they are aligned along the direction of the local magnetic field. An auroral observer who forgets this fact sees mass confusion when his sky is filled with aurora. The awesomeness of the scene may be quite enough to make the observer happy, but it is more satisfying to be able to visualize mentally the true shapes of the auroral structures the eye sees.

Recall that in central and northern Alaska the earth's magnetic field is nearly vertical (differing by only 10 to 15 degrees). Hence, discrete auroral forms seen there are structures that stand nearly vertically in the sky. Each discrete auroral form extends upward above its lower border precisely along the direction of the magnetic field. The lower border lies parallel to the earth's surface and is typically located 80 to 120 km above. From the lower border, the form extends upward several tens of kilometers, and sometimes for several hundred. Discrete auroras also are elongated horizontally, most usually in the magnetic east-west direction. They may stretch for several thousand kilometers, far beyond what an observer at one location can see. Yet despite their great length and their considerable height, most discrete auroras are incredibly thin. Their thickness may be no more than 100 meters, the length of a football field. The typical discrete auroral form is indeed a curious structure: long, tall in the vertical direction, yet very thin, much like a wide ribbon used to wrap gift packages, and sometimes of similar color.

Sec. 5.2.1 Names Applied to Discrete Auroral Forms

A discrete auroral structure that exhibits slight and uniform curvature along its length and extends from horizon to horizon is called an *arc*. Plate 3 contains an excellent example of multiple auroral arcs. If the structure has somewhat irregular curvature, as in Plate 2, auroral observers are inclined to call it a *band*. Segments of arcs and bands (other portions are perhaps too weak to be seen) might get called *draperies*. The rayed structure in Figure 5.1 could be called a drapery. Figure 5.2 contains both an arc and a band,

Figure 5.1 An unusually good photograph of a rayed drapery, or portion of an arc or band—too rarely do the rays remain stationary long enough for such photographs. (V. P. Hessler photograph.)

and Figure 5.3 shows a highly curved band. Remember that these names—arc, band and drapery—all refer to one basic kind of auroral structure, although each name says something about the shape of that structure. Over the course of a few minutes, an arc might develop enough extra curvature

Figure 5.2 A rayed auroral arc high in the sky, with a band beyond it, nearer to the horizon. (V. P. Hessler photograph.)

to deserve the name band (just when is up to the individual observer). Later on it might fold on itself to form long subparallel arms such as shown in the all-sky camera[1] photograph in Figure 5.4. It might also curl up on itself so much that it is rightly called a *spiral*, the most convoluted form of all. (See Figure 5.5 and the closest aurora in Plates 5 and 6.)

1. An all-sky camera has a fish-eye lens 180° wide. Its images show the the horizon around the edge of the pictures, and the geographic (true) zenith at center.

Figure 5.3 A highly curved band with ray structure, at about 20 degrees above the north horizon. Other aurora lies beyond. (V. P. Hessler photograph.)

At a location such as Fairbanks, which lies just south of the auroral zone, it is common during the evening hours to observe two or more parallel and closely spaced arcs or bands in the northern sky. If these become rayed and also move southerly, coming overhead so as to straddle the magnetic zenith, they may create a spectacle that observers refer to as the *corona*. (See Figure 5.6 and Plate 13.) If an observer reports seeing a corona, anyone receiving the report knows exactly where the aurora was because corona can occur only directly over the observer's location, straddling his magnetic zenith.

Sec. 5.2.2 Homogeneous and Rayed Discrete Auroras

Homogeneous arcs and bands are those that exhibit little or no internal structure. *Rayed* arcs and bands contain thin striations called *rays*. The rays are aligned exactly along the direction of the earth's magnetic field, hence they are nearly vertical over Alaska, with the

Figure 5.4 All-sky photograph of a band in the zenith over Fort Yukon, Alaska, that contains very sharp elongated folds. Other arc-like forms lie both to the north and south.

upper tips slightly south of the lower ends. A rayed auroral arc seen low in the sky looks somewhat like a picket fence. Rayed discrete auroras tend to be brighter than homogeneous ones.

Sec. 5.2.3 Auroral Rays

Auroral *rays* form when a thin auroral arc curls up on itself to create tight bundles that appear much brighter than the intervening portions of the arc between the rays. The apparent increased brightness of the ray occurs because observers looking at a ray are seeing several layers of the arc, and the light coming from one layer adds to the others. Auroral rays are small, only a kilometer or so in diameter. They form quickly and also tend to move horizontally along the rayed auroral form. For these reasons, viewers are unable to recognize rays when looking at them end-on (in the viewer's magnetic zenith). But the use of highly sensitive television cameras for auroral observation has shown that the rays are as described here, tight bundles of light also called *curls*. Figure 5.7 contains three images of ray structures acquired with television cameras. The top image shows rays seen end-on in the magnetic zenith, portraying well their curled shape. The rayed structure in the middle panel is a few degrees north from the magnetic zenith, and that in the bottom panel is approximately 30 degrees north of the zenith. The two lower panels show the same aurora captured at two different times, the

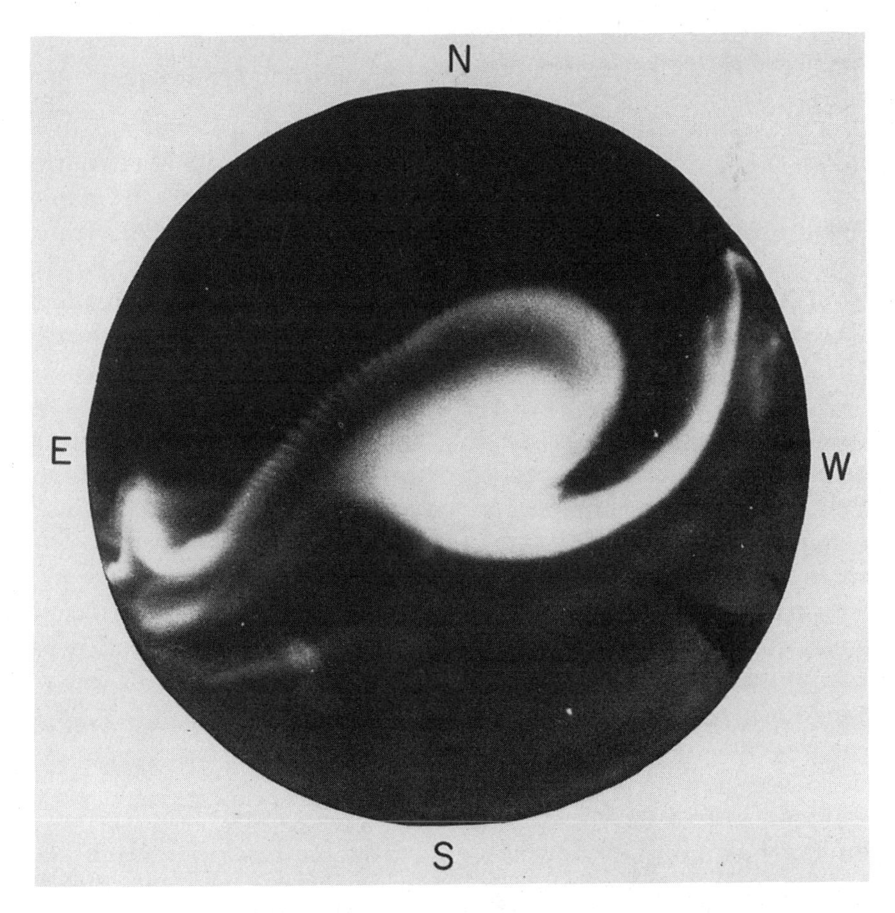

Figure 5.5 A spiral (also called a surge) in the zenith over Fort Yukon, Alaska. Note the rays in the arm of the spiral and, off to the east, another spiral-like convolution. The dark object at lower right is a spruce tree.

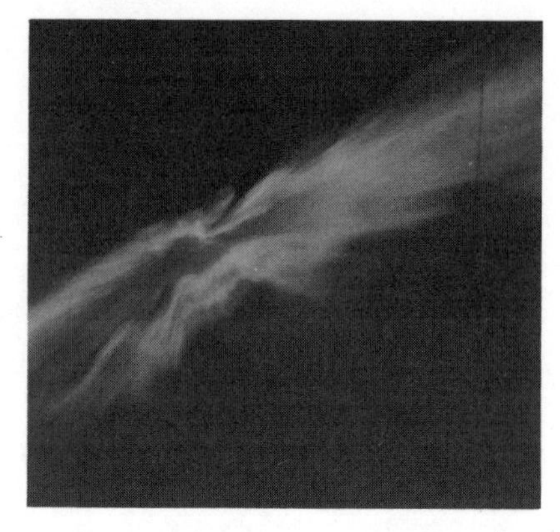

Figure 5.6 Examples of corona. (Left-hand photograph by Carl Störmer; right-hand photograph by V. P. Hessler.)

bottom image first, then the middle image several minutes later as the rayed aurora moved toward the magnetic zenith.

Section 5.3 PULSATING AURORA

Pulsating auroral forms are those that undergo periodic or quasi-periodic variations in brightness with periodicity ranging from about 0.1 sec to somewhat more than 20 sec. Pulsating auroras are always weak and therefore difficult to recognize. Because of their weakness, they usually appear colorless.

To see pulsating aurora, a watcher should look overhead during the minutes or hours following the appearance and subsequent disappearance of discrete aurora (in other words, after a breakup), making sure that your eyes are well dark-adapted. The aurora probably will look patchy at this time. Fasten the eyes on one patch. Within a few seconds, that patch will probably disappear, but it will reappear a few seconds later. Once a watcher of auroras has recognized a pulsating auroral form, he or she will have no trouble seeing others. A look to the east and west of the one seen will reveal many more. Once pulsating auroras occur in the sky, they may continue for hours, perhaps until dawn.

Figure 5.7 (at right) Three views of auroral rays: At top, the rays seen end-on, displaying their 'curl' form. At center, rays slightly out of the magnetic zenith and, at bottom, the same rays seen displaced from the north of the zenith by approximately 35 degrees.

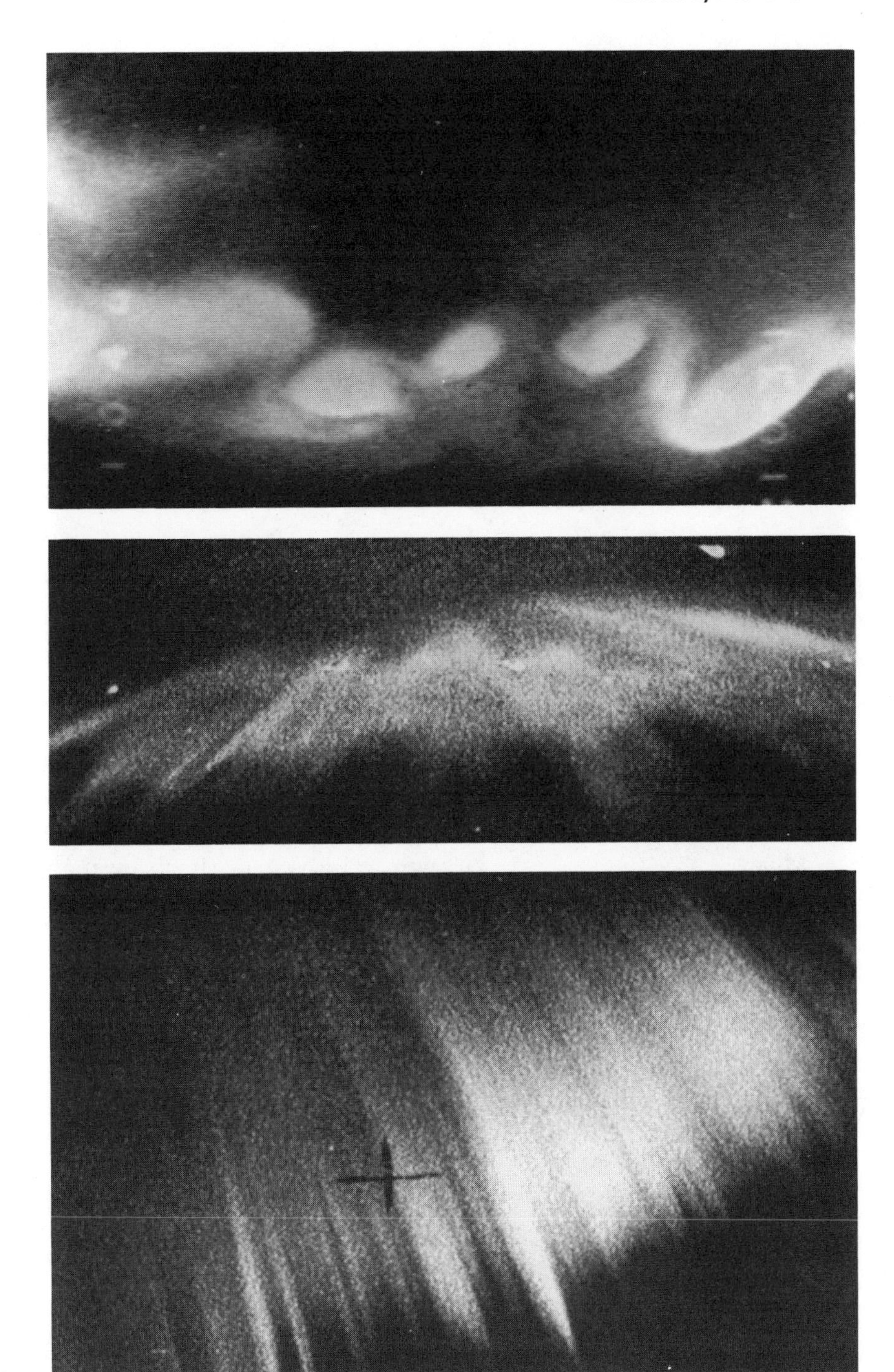

Figure 5.8 shows two sets of television images of pulsating auroras. Close inspection will reveal significant differences within the images of each set. The structures seen here are so small that a visual observer probably would not recognize them. When an observer does recognize pulsating aurora, it is more likely to look like the collection of patchy structures seen in the sequence of all-sky photographs shown in Figure 5.9. Figure 5.10 contains tracings that show the temporal variations within pulsating auroras. Note in the fourth panel the regular nature of the variation as compared to that in other panels, where the brightness changes are more irregular and burst-like. The variety of brightness alterations in pulsating aurora is endless.

Pulsating aurora is spectacular, but it lacks the brightness, color and fast motions typical of the discrete aurora. In the overall scheme of things, pulsating aurora is important because it is widespread and therefore represents the end effect of a substantial portion of the energy carried into the global auroral atmosphere by incoming fast particles. Some observations suggest that the 'on' phase of a pulse is associated with an increase in energy of the responsible incoming particles because the altitude of the lower border during the 'on' phase is lower than the altitude of nearby diffuse nonpulsating background aurora that may accompany the pulsating forms. Observations show that, unlike the discrete aurora, pulsating aurora sometimes has limited height extent; that is, it may occur essentially in horizontal layers (See Section 5.5.4, below).

Section 5.4
THE HYDROGEN ARC
(THE DIFFUSE AURORA)

If discrete auroras are observed in the northern sky or overhead at a location such as Fairbanks during the evening hours, then the aurora watcher can usually see the hydrogen arc. The *hydrogen* (or *proton*) arc is a visible manifestation of the impingement on the atmosphere of a fast-particle stream that is especially rich in protons, although electron bombardment actually creates most of the brightness. The cause of this relatively featureless type of aurora is a steady drizzle of electrons and protons into the atmosphere from the outer portion of the Van Allen belt (See Section 13.1). The hydrogen arc

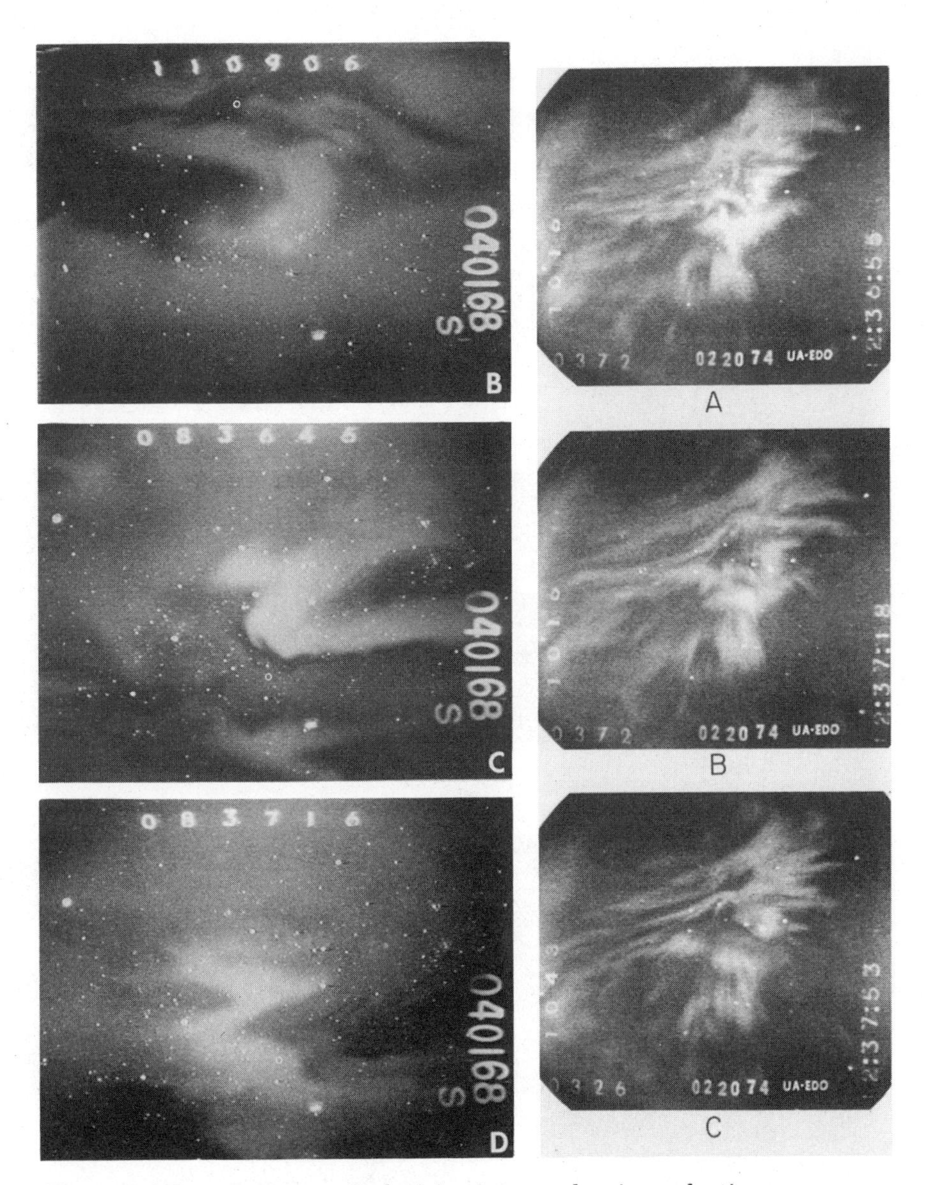

Figure 5.8 Two sequences of television images showing pulsating auroras. In the sequence at the left the location of the magnetic zenith is shown by a white circle. The sequence at the right contains highly structured pulsating forms, much smaller than those at left. The images in each sequence are about 30 seconds apart.

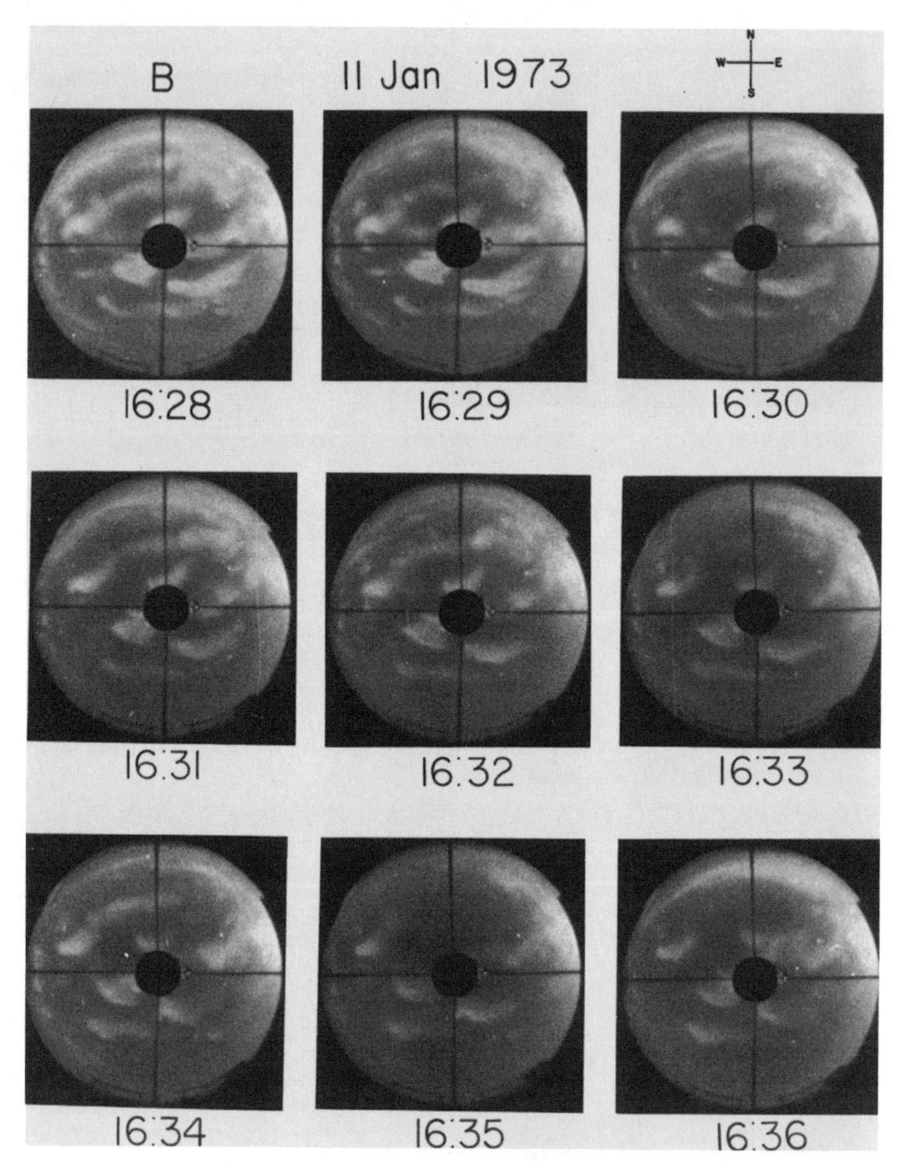

Figure 5.9 All-sky photographs of pulsating patches taken at intervals of 1 minute.

is a highly uniform, broad band of light extended in the east-west direction that typically lies southward of and separated from the discrete portion of the display. Often no brighter than the Milky Way and similar to it in appearance, the hydrogen arc is not easy to recognize or photograph. Compared to the dis-

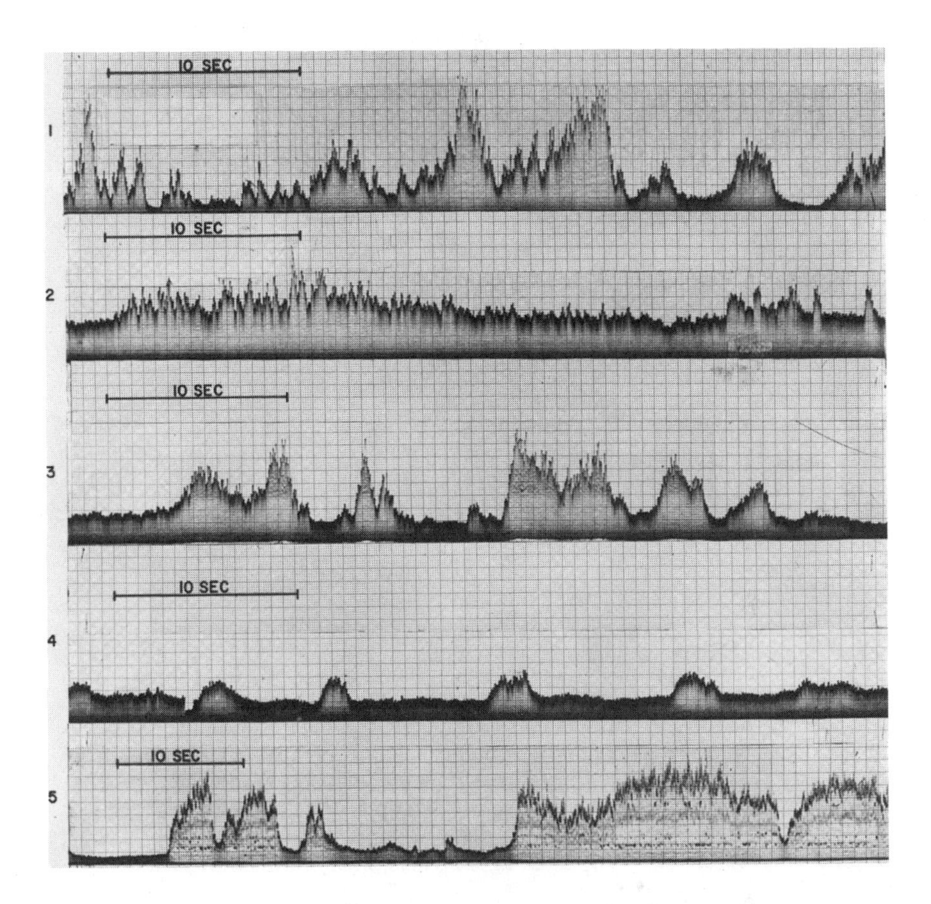

Figure 5.10 Five examples of intensity variations of pulsating aurora, obtained by a recording photometer pointed at a television screen on which the tape-recorded images were played back.

crete and pulsating auroras, it is visually uninteresting. If the arc is bright, the observer may recognize some internal structure, but often the hydrogen arc appears featureless and, because it is so weak, colorless. The spatial relationship of the hydrogen arc to the rest of the auroral display is best shown in satellite imagery (see the example in Figure 5.11). The name 'diffuse aurora' is increasingly being attached to the hydrogen arc.

Section 5.5 OTHER KINDS OF AURORA, OR SPECIAL CONDITIONS

On infrequent or rare occasion, an observer may see curious auroral phenomena not

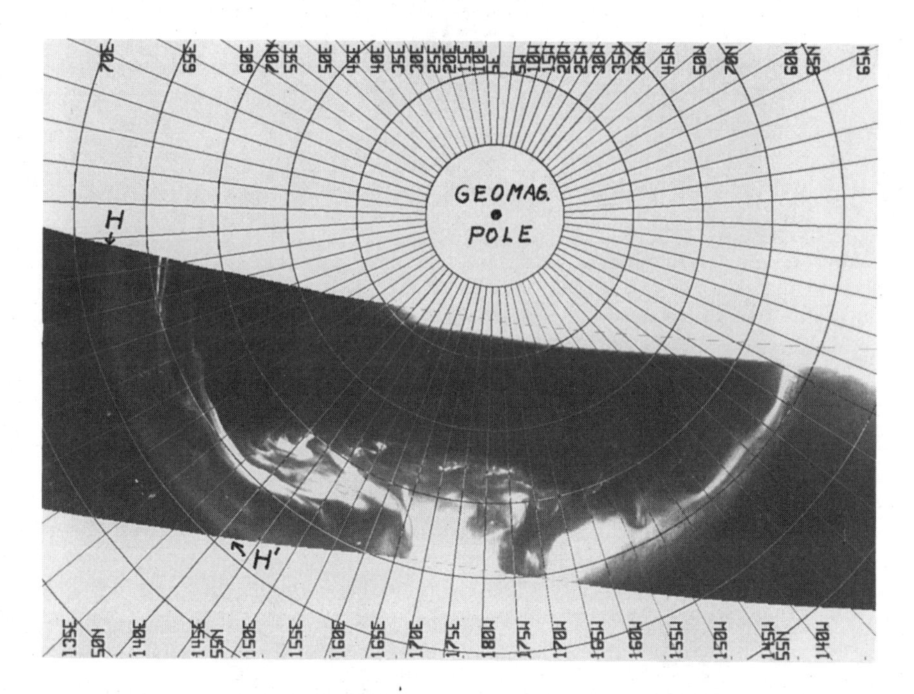

Figure 5.11 A satellite view of a large portion of an auroral display superimposed on a grid showing geomagnetic latitude and longitude. The direction to the sun is toward the top. The hydrogen arc appears as a weak, featureless band near the left of the image, between points H and H'. The bright aurora to the north and east of it is discrete aurora, and that farthest right is pulsating aurora. City lights in the Soviet Union show at lower left.

described above. The exception is *enhanced aurora*, which occurs more frequently, perhaps at least once on any long night of observation.

Sec. 5.5.1 Enhanced Aurora

Enhanced aurora exhibits abnormal brightness in one or more thin horizontal layers lying at or not far above the aurora's lower border, as shown in Figure 5.12. The characteristics of the bright layers suggest that they are caused by an abnormal interaction between the incoming particle streams and the resident atmospheric gases. This extra interaction probably is one physicists call beam-plasma discharge. In a complex fashion, it draws extra energy from the incoming particles and passes it on to the atmospheric gases. Its end result is all that need concern us here: thin layers of enhanced auroral brightness that sometimes lend added beauty to twirling auroral skirts by giving them extra hems.

Sec. 5.5.2 Flickering Aurora

Not a form type as such, *flickering* is a condition pertaining to bright discrete auroras wherein internal striations (ray-like structures) appear to vibrate or flicker in the fashion of a candle's flame exposed to a gentle draft. The flickering fluctuations seem to be both spatial and temporal, and photometric measurements of flickering indicate periodic changes with a frequency in the range 7 to 10 Hertz (cycles per sec). Instrumental observations show that this phenomenon occurs frequently for a few minutes during the times when discrete auroras are at their brightest—near the time of the auroral breakup. However, visual observers see it only rarely, when the flickering is especially intense.

Sec. 5.5.3 Flaming Aurora

Flaming aurora appears to have repeated waves of enhanced brightness moving up along the forms from their lower

Figure 5.12 Enhanced aurora, characterized by one or more bands of unusual brightness above and parallel to the lower border of rayed arc. Television photograph courtesy of T. J. Hallinan.

borders. The name comes about because the effect looks similar to flames in a wood fire. Observers usually see flaming aurora when broad regions of pulsating aurora lie either to the north or south of their observing locations. Scientists do not know if the flaming appearance is inherent to the normal pulsating behavior or if it signifies a special condition affecting only some pulsating forms.

Sec. 5.5.4 Black Aurora

Black aurora is not aurora; rather it is an absence of aurora in small bean-shaped pods or thin arc-like regions surrounded by widespread uniform aurora within the pulsating part of the display. The surrounding aurora does not pulsate, and the fact that the black voids do not change shape as they drift across the sky (typically eastward) implies that the surrounding aurora is in the form of horizontal sheets rather than being aligned along the direction of the magnetic field like the discrete aurora. 'Black' aurora is unexplained, but certain instrumental observations indicate that the black regions possess excess positive electrical charge (as contrasted to auroral rays, which are centers of excess negative charge.) Aurora observers see this phenomenon rarely, and then typically only late in the night. Figure 5.13 contains examples of 'black' aurora obtained with television cameras.

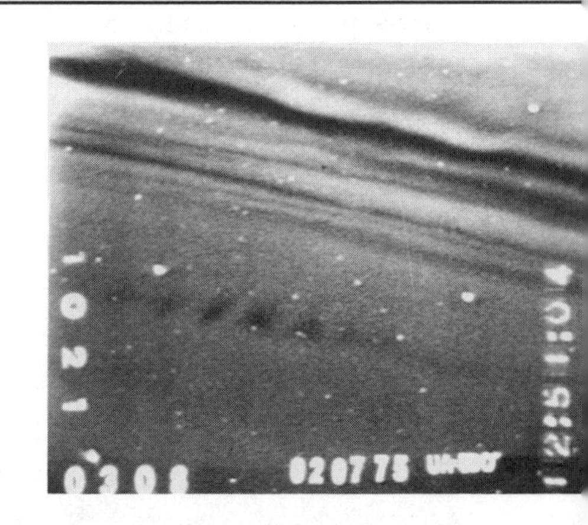

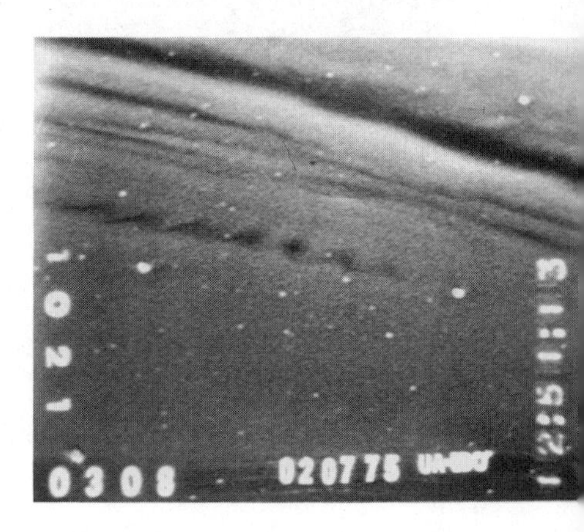

Figure 5.13 (at right) Television images of 'black aurora,' narrow-field images at left and right, and all-sky images at center. The right-hand pair of images show, except for a dark streak of 'black aurora,' a uniform veil-like glow of aurora covering the field of view of the television camera.

Sec. 5.5.5 Fast Auroral Waves

Fast auroral waves are southward-travelling enhancements of pre-existing auroral forms within the pulsating part of moderate to large auroral displays. The waves look remarkably like an airport beacon light repeatedly sweeping southward across the sky from a source located far to the east or west of the observer. Observers located well to the south of the auroral zone are more likely to see the waves than are those close by. The fast auroral waves typically occur late in the night. Their cause is unknown.

Sec. 5.5.6 Veil

Veil is an extremely rare phenomenon involving the appearance of a broad band of featureless auroral emission

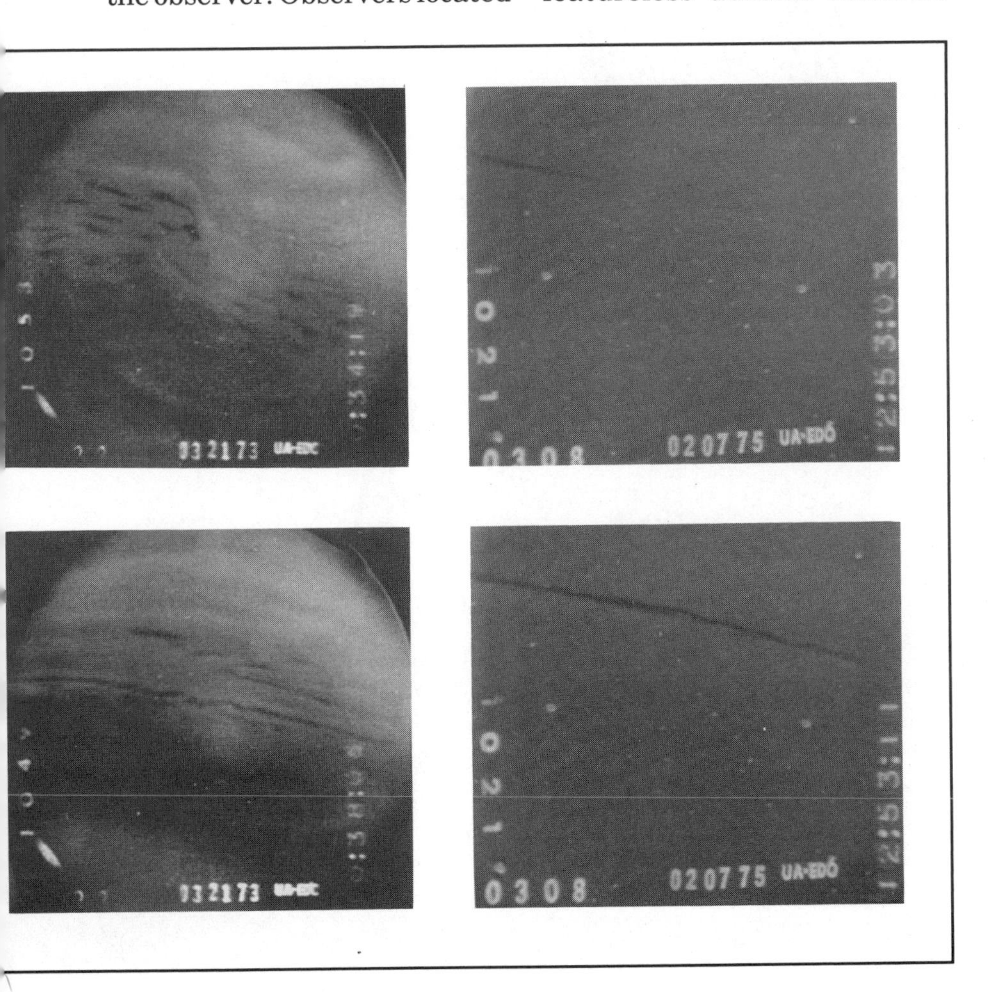

An artist's rendition.

aligned in the east-west direction, but which may be wide enough to cover an observer's entire sky. A veil may be brighter than either the hydrogen arc or the pulsating aurora, and, on rare occasion, much brighter. The broad expanse of aurora shown in the two right-hand panels of Figure 5.13 is typical of the appearance of veil aurora.

Section 6
TEMPORAL VARIATIONS IN THE AURORA

Section 6.1
THE RANGE OF TEMPORAL VARIATIONS

The temporal variations of immediate interest to aurora watchers are those of relatively short duration, the ones that determine if an aurora might be seen on a given night, when during the night, and how interesting the aurora will appear at any moment. The full range of recognized variations is broad, extending from at least centuries down to a fraction of a second.

Section 6.2
EXTREMELY LONG-TERM VARIATIONS

Both long-term variations on the sun and changes in the earth's magnetic field during past geologic eras probably have had profound effects on the aurora, but few details are known. Some observations indicate that the auroral zones have not always been where they are now, because the earth's magnetic field has changed with time.

Section 6.3
LONG-TERM VARIATIONS

Any change on the sun that causes variation in the flow of charged particles (electrons and protons) toward the earth will affect the aurora. Times of strong flow produce the most intense and widespread auroras. Study of the historical record suggests that relevant solar changes do occur with periodicities of about 80 years and 250 years. Reported approximate periods of maximum activity since the time of Christ are: 1st Century A.D. and years near 350, 500, 650, 1150, 1375, 1610, and 1870. Sometime about 2010 is predicted to be another period when the long-term cycle peaks. The periods of recognized minimum activity are those near 425 A.D., 680, 1000, 1265, 1475, 1680 and 1740. This long-term cycling also called for a minimum in about 1990. However, shorter-term variations of types discussed below, especially the 11-year solar cycle, can overpower the effects of the 80- and 250-year cycles, and it so happens that 1990 fell on a maximum of the 11-year cycle and was a very active year. Similarly, the period near 2010, predicted to be at a maximum of the long-term cycle, may actually prove to be less active than decades preceding and following.

During one well-documented low period, the Maunder Minimum (from 1645 to 1715), the sun was very quiet and few auroras were observed at middle latitudes. A great auroral display in 1716—more extensive than any during the previous 142 years—terminated the Maunder Minimum. It terrorized many residents of Europe and greatly titillated the scientists of the time. England's famous astronomer Edmund Halley was then 60 years old. He had been observing the sky with care for many years, but this was the first aurora he ever saw.

It also was the first for Anders Celsius, the inventor of the centigrade thermometer scale. Then only 15 years old, Celsius began compiling a catalog of auroral displays, and the 316 entries he made during the next 16 years certainly proved that the Maunder Minimum was over.

Section 6.4
11-YEAR SOLAR CYCLE VARIATIONS

Although auroras can be seen at the auroral zone on any dark, clear night, they tend to be brighter and more extensive during the maximum years of the well-known 11-year sunspot cycle. During those maximum years, some auroral displays are more likely to extend down to middle latitudes, and a few great displays might occur. The solar cycle is not precise; peaks occur every 7 to 15 years, but the average period is 11 years. (See Figure 6.1.) Solar maxima should occur in about 2002, 2013 and 2024, but the actual peaks may be several years before or after these dates.

The reason for the 11-year cycle in auroral activity has to do with the location of active

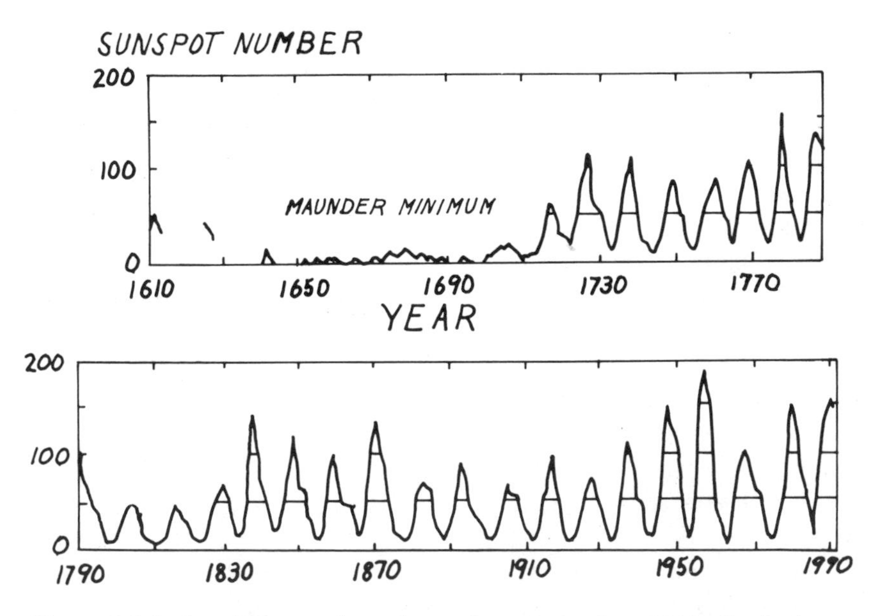

Figure 6.1 A plot of observed numbers of sunspots since 1610, displaying the 11-year solar cycle from about 1700 onward.

regions on the sun's disk as seen from the earth. These active regions typically are in the vicinity of sunspots and they may contain prominences (see Plate 20) and solar flares (see Plate 21). The active regions emit impulsively enhanced flows of the particles that generate the aurora, so when these regions are near the center of the sun's disk as seen from the earth they spray particles out toward the earth. Solar flares occurring within the active regions carry the material out away from the sun, whereas the prominences tend to carry ejected material back to the sun's surface, thereby suppressing the solar wind. Over the course of an 11-year solar cycle (going from solar minimum to solar maximum), sunspots and their associated flares and prominences first are most prevalent at high solar latitude, and then move toward the solar equator. Only when the sunspots are near the equator do the associated flares shower the earth with the enhanced flows of ejected particles. But even when sunspots and flares are absent, the sun boils off enough particles from other parts of its surface to create continuous aurora over the earth's polar regions.

Section 6.5
SEASONAL VARIATIONS

Near the auroral zone, the main factor determining the number of auroras seen throughout the year is the amount of darkness. At lower latitudes, however, a spring and fall maximum is noticeable. The reason is thought to be that in these seasons the earth is farthest north or south of the sun's equator. Therefore, in spring and fall, the earth is more likely to intercept enhancements of the solar wind that emanate from the vicinity of sunspots.

Section 6.6
MAGNETIC STORM VARIATIONS AND THE 27-DAY RECURRENCE TENDENCY

The term 'magnetic storm' refers to intervals of one to several days during which an enhanced flow of solar particles strikes the earth. Among the observed effects are changes in the strength and direction of the magnetic field observed at the earth's surface and enhanced displays of aurora. If a pre-existing large sunspot approaches the center of the sun's disk, a magnetic storm is likely to occur. Also, a flare may suddenly erupt in that position and eject material toward the earth. Active regions associated with large sunspots may persist for several months, and the sun, as seen from the earth, appears to rotate on its axis with a period of approximately 27 days. Hence, an active region that generates a magnetic storm on one date may generate another 27 days later. (See box, page 69.)

Section 6.7
SUBSTORM VARIATIONS

An auroral substorm is an important event, both for the casual watchers of auroras and for the scientists who are seeking to understand all the ramifications of the solar-terrestrial interaction. Substorms involve almost explosive activations of the aurora in a period of a few to a few tens of minutes, and the overall duration of a substorm is about 1 to 3 hours. The auroral substorm is a global phenomenon that affects all of the

Figure 6.2 (at right) The progression of an auroral substorm depicted as seen from high above the geomaagnetic pole. The diagrams illustrate how the substorm starts near the midnight meridian and expands east and west, and also poleward and equatorward. Panels E and F illustrate the recovery phase of the substorm. (Diagram by S.-I. Akasofu.)

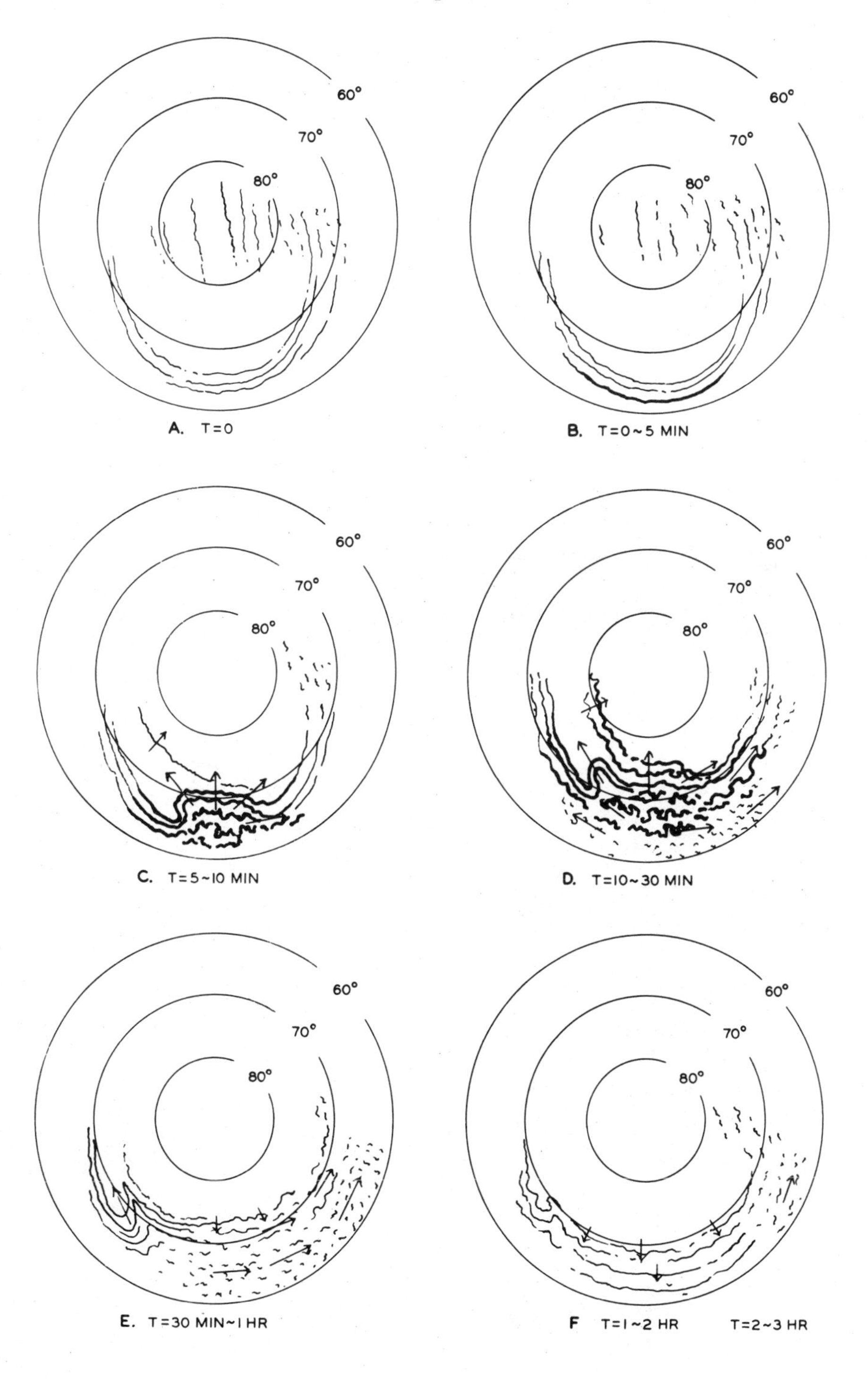
60°
70°
80°
A. T=0
60°
70°
80°
B. T=0~5 MIN
60°
70°
80°
C. T=5~10 MIN
60°
70°
80°
D. T=10~30 MIN
60°
70°
80°
E. T=30 MIN~1 HR
60°
70°
80°
F T=1~2 HR T=2~3 HR

individual auroral forms within the auroral oval as well as the shape of the oval itself. A moderate to large substorm usually can be recognized as consisting of three parts: growth phase, expansive phase and recovery phase. During the growth phase the auroral oval expands in the region near the midnight meridian so that it is seen to move southward (in the northern hemisphere), and the auroral forms contained in the oval may show an increase in brightness. The following expansive and recovery phases are depicted in Figure 6.2. The onset of the expansive phase, shown in part B of Figure 6.2, typically involves a brightening of the southernmost auroral form in the oval. The brightening usually commences near the midnight meridian and expands both eastward and westward along the oval (see parts C and D of Figure 6.2, and also Figure 6.3),

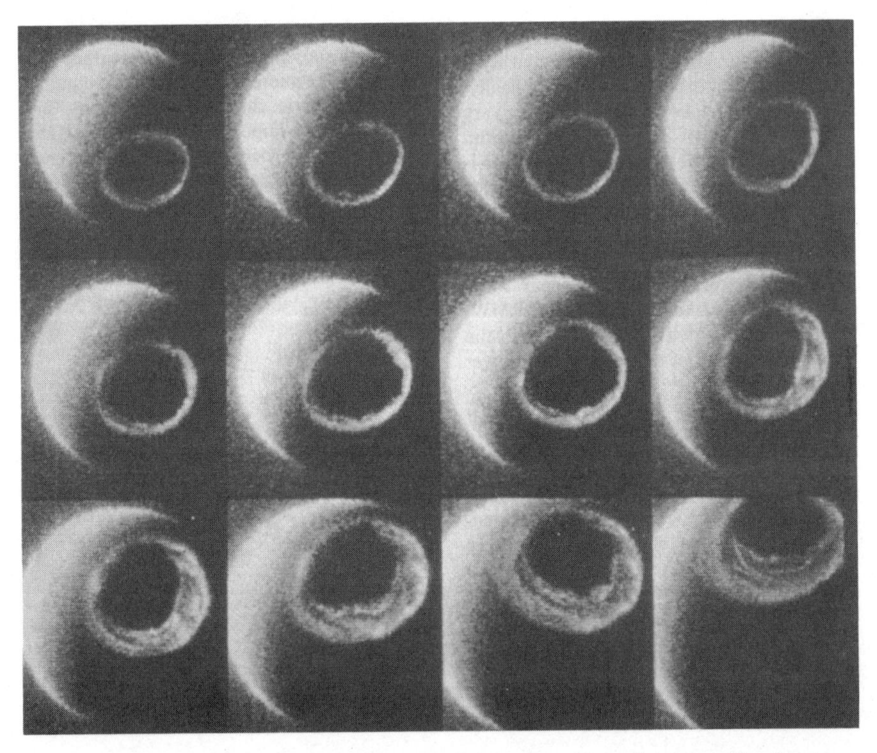

Figure 6.3 Successive satellite images of the auroral oval taken 12 minutes apart, showing the changes to the oval during the development of an auroral substorm. (Courtesy of J. D. Craven and L. A. Frank, University of Iowa.)

reaching all the way around the earth to the noon meridian. Seen near the midnight hours, this is the event referred to as the auroral breakup. During the subsequent recovery phase, parts E and F of Figure 6.2, patchy pulsating aurora usually predominates.

During times of magnetic quiet, when only a comparatively steady and not intense flow of particles is coming to the earth from the sun, auroral substorms occur several times during each 24-hour period. These substorms are noticeable but small: the aurora typically brightens somewhat for a few minutes but otherwise does not become overly spectacular. Moderate to large auroral substorms occur during magnetic storms. Each individual substorm tends to last longer then, and they are more frequent, perhaps so frequent that they essentially overlap. The auroral oval typically broadens and expands under the impact of these intense substorms.

What an aurora watcher at the auroral zone sees when a moderate to large auroral substorm occurs depends on the watcher's location with respect to the midnight meridian, that is, whether it is early evening, near midnight, or after midnight at the observer's location. (The observer's longitude also is a factor. Because of the offset of the geomagnetic pole from the geographic pole, the observed sequence in eastern Canada occurs an hour or two earlier in local time than in Alaska, and in Scandinavia several hours earlier.)

Evening hours: The observer is likely to see a multiple array of discrete auroras (arcs and bands) and perhaps the weaker hydrogen arc (the diffuse aurora) lying south of the region of discrete aurora. The discrete auroral forms brighten as the substorm proceeds, and homogeneous forms tend to become rayed. The arcs and bands may undergo contortions that convert them to spiral-like forms (also sometimes called westward-travelling surges because they may move rapidly westward). This is the breakup, in past sometimes called the pseudo-breakup because it is usually not as spectacular as the breakup seen nearer midnight. Usually, but not always, the contortions and brightenings proceed from the east toward the west along the length of the forms. Observers may see rapid motions of rays or larger contortions along the forms, both toward the east and west, but they are likely to receive the impression that westward motions predominate. The overall display likely will become wider in the north-south direction.

Near-midnight hours: During the growth phase of the substorm the oval expands and moves southward. Then as the expansive phase begins, the observer near the midnight meridian will usually see a rapid brightening of the southernmost of the discrete auroral forms. Within seconds, the brightness of this arc or band can increase several fold. Almost simultaneously, the brightening form will undergo severe distortion, and it and any nearby auroral forms usually develop into a spiral-like configuration that might fill much of the observer's sky. The change is so rapid and the scene so awesome that all observers quickly recognize that a breakup is under way. After a few minutes of violent activity, the display typically weakens and leaves remnants of the previously well-ordered auroral configurations scattered across the sky. A transition has occurred that usually signals the commencement of the pulsating portion of the display. The substorm is now said to be in its recovery phase.

Postmidnight hours: During the hours after midnight, an observer near the auroral zone is most likely to be watching pulsating aurora. The individual forms usually drift slowly eastward. If the forms are patchy, and then begin to reform into arc-like structures aligned east-west, a new breakup probably will occur. If it does, the brightening and contorting of forms probably will progress from west to east and take place somewhat more gradually than during a breakup observed near midnight. The observer also might see nothing more than a general increase in the brightness and extent of pulsating aurora.

Daytime Hours: Observers located in North America have no opportunity to see the aurora occurring on the dayside of the earth. Were they in Svalbard (lying between northern Greenland and Scandinavia) or in parts of Antarctica, they could, because at these locations midwinter darkness allows auroras to be seen even at noon. Observers favorably located would see a general increase in auroral activity, typically of pulsating aurora prior to noon, and of arc-like discrete auroras in the afternoon.

Section 6.8 SHORT-TERM VARIATIONS

The fastest recognized variations in the aurora are the periodic and quasi-periodic changes in intensity that occur in flickering and pulsating auroras, down to 0.1 sec. The repeating nature of these variations suggests that their cause is related to interactions between incoming particle streams and electromagnetic waves that exist in the magnetosphere and the ionosphere. In addition, the person who watches aurora for any length of time is likely to get the impression that the brightness of auroras and the general level of activity vary, perhaps in irregular fashion, over periods of several minutes. The reason for this variation is not known.

Great Magnetic Storms of Recent Times

During these events, extensive auroral displays occurred over large parts of the earth's surface.

Month and Day	Year
August 28-September 7	1859
February 4	1872
November 17-21	1882
October 31-November 1	1903
September 25	1909
May 13-16	1921
April 16	1938
February 11	1958
July 8	1958
August 4	1972
December 19	1980
March 13, 14	1989

COLOR PLATES

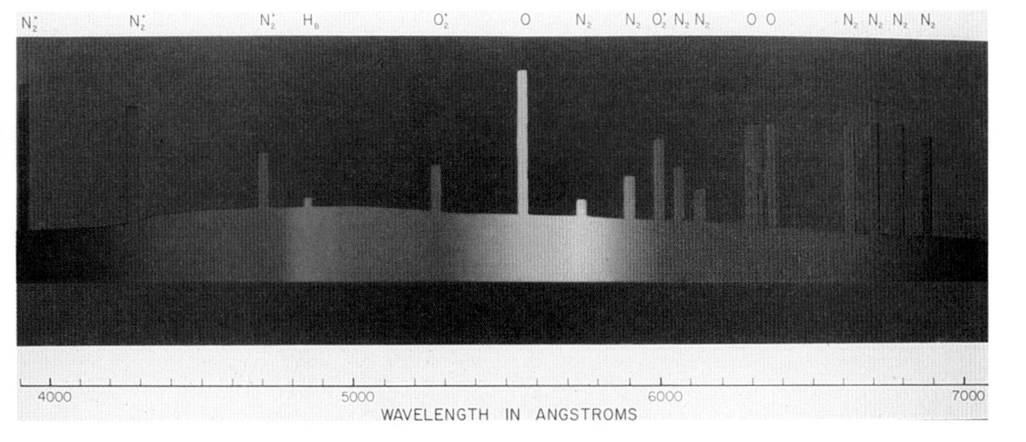

Plate 1. *The most prominent lines and bands in the visible auroral spectrum, their intensity represented by the color strips standing above the solar spectrum shown below.*

Plate 2. *Green auroral bands with wavy structure; color Type C.*

Plate 3. *A remarkable array of parallel arcs over the University of Alaska Fairbanks campus, color Type C. (Takeshi Ohtake photograph.)*

Plate 4. *Green bands showing blue hue in the photograph; color Type C. (Gustav Lamprecht photograph.)*

Plate 5. *NASA Shuttle photograph showing a color Type C aurora of the southern hemisphere in foreground, with color Type A in the distance; taken in 1991. (NASA photograph courtesy of T. J. Hallinan and Don Lind.)*

Plate 6. *Same as Plate 5, but taken some seconds later. (NASA photograph courtesy of T. J. Hallinan and Don Lind.)*

Plate 7. *Color Type C green bands over the Fairbanks area. (Takeshi Ohtake photograph.)*

Plate 8. *Color Type C green bands showing a blue hue, because of moonlight. (Gustav Lamprecht photograph.)*

Plate 9. *Rayed bands, color Type C, seen through a communications antenna. (Robert Hunsucker photograph.)*

Plate 10. *Green bands with red upper parts, color Type A, photographed over the Brooks Range. (David C. Fritts photograph.)*

Plate 11. *Type A aurora of the southern hemisphere photographed from a NASA Shuttle in 1991. The glow around the tail of the Shuttle is emission from attitude-control jets. (NASA photograph courtesy of T. J. Hallinan and Don Lind.)*

Plate 12. *Uniform Type A aurora of the southern hemisphere photographed from a NASA Shuttle in 1991. (NASA photograph courtesy of T. J. Hallinan and Don Lind.)*

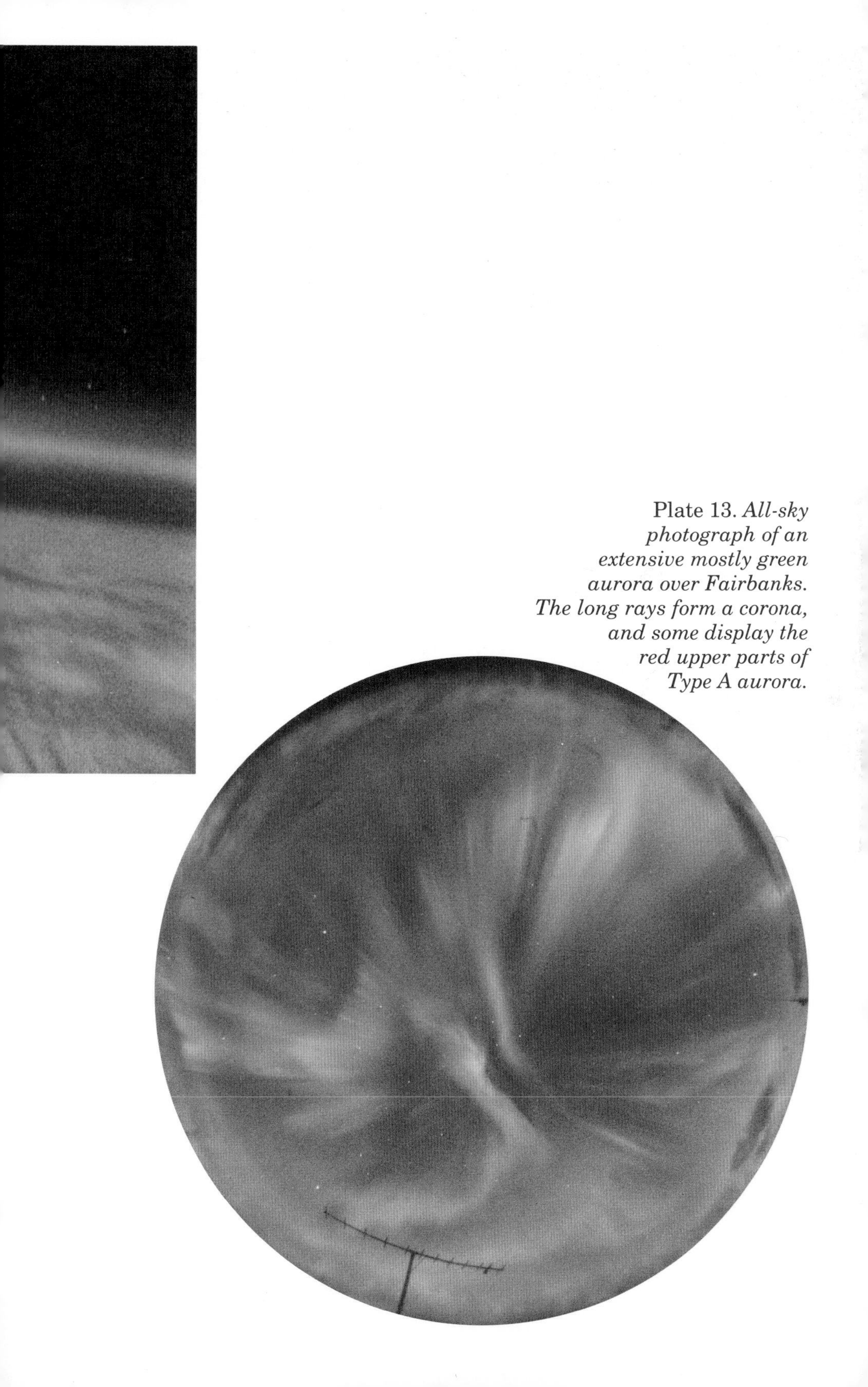

Plate 13. *All-sky photograph of an extensive mostly green aurora over Fairbanks. The long rays form a corona, and some display the red upper parts of Type A aurora.*

Plate 14. *Mostly Type C green bands, but showing slight hints of Type A red in the upper portions.*
The photographer built the house and cache in foreground.
(Gustav Lamprecht photograph.)

Plate 15. *Rayed Type C and Type A bands in moonlight. Note the low clouds that hide part of the display. (Gustav Lamprecht photograph.)*

Plate 16. *Auroral arc with a pronounced red lower border; color Type B.*

Plate 17. *Green Type C and Type B (red lower border) bands over the Brooks Range. (David C. Fritts photograph.)*

Plate 18. *Type D all-red aurora photographed through a birch tree in the author's yard in 1959.*

Plate 19. *As in Plate 18, but looking directly south at all-red aurora.*

Plate 20. *A large flare prominence looping up above the sun's surface. Note its striated nature, much like that of the aurora. The two feet of the prominence are within sunspots.*

Plate 21. *Another eruptive flare as in Plate 20; again note its striated nature.*

Plate 22. *Simultaneous photographs showing exactly conjugate auroras in the two hemispheres, well illustrating the mirror-imaging of the northern and southern hemisphere auroras. Taken on March 26, 1968.*

Plate 23. *A ground-test firing of a barium release canister.*

Southern Hemisphere

Plate 24. *A barium release photographed from the ground several minutes after release. The purple ion cloud shows elongation and striations, whereas the green neutral cloud remains roughly spherical.*

Plate 25. *A long-exposure photograph (note star tracks) during a rocket flight from Poker Flat, Alaska. The intense part of the bright arch at lower left is caused by the firing of the first-stage rocket motor. The hot motor glows enough to record the stage's fall back to the ground. The streak above is caused by the firing of the second stage. This rocket system carried two barium shaped-charges fired at altitude 800 km on the up- and down-legs of the trajectory, high over and north of the auroral arcs seen crossing the center of the photograph. (Photograph by Greg Boquist.)*

Plate 26. *A color Type A rayed auroral band partly hidden by low clouds. Photograph taken by C. Russell Philbrick at Poker Flat, looking south toward Fairbanks where the city lights illuminate the clouds overhead.*

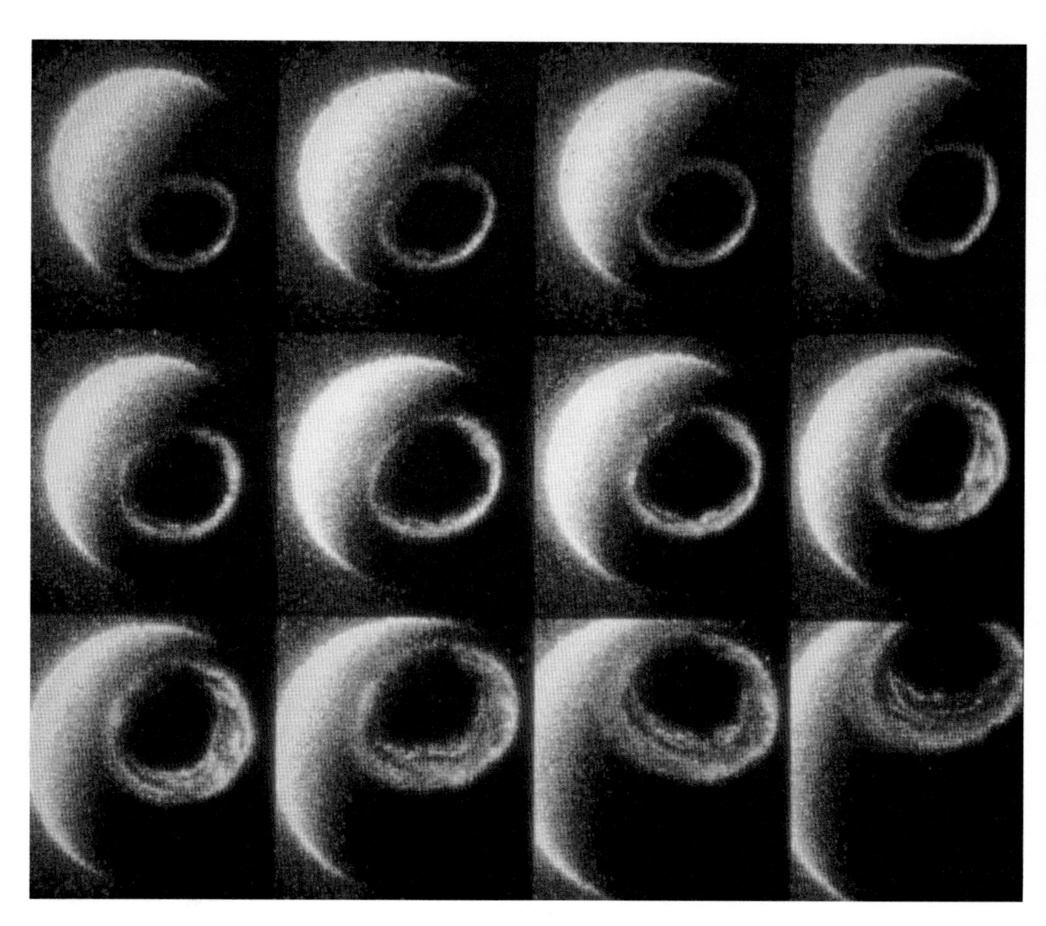

Plate 27. *A color view of Figure 6.3, containing successive images of the auroral oval taken 12 min apart, showing the changes to the oval during the development of an auroral substorm. (Courtesy of J. Craven and L. Frank, University of Iowa.)*

Part II
FOR THOSE WHO WANT TO LEARN MORE

Section 7
AURORAL COLORS AND THEIR CAUSES

Previously, in Section 4.7, the main patterns of auroral coloration were listed:

Type A Green auroras shading to red at their tops.

Type B Green auroras with red lower borders.

Type C Auroras green throughout.

Type D All-red auroras.

Type E Fast-moving Type B auroras.

Type F Blue-purple sunlit auroras.

Four key factors govern the color of an auroral form:

1. The energy of the incoming particle beam.

2. The altitude-dependent distribution of the various atoms and molecules in the atmosphere.

3. The characteristics of the atoms and molecules in the atmosphere that cause each species to emit auroral light of different colors.

4. The increasing density of gases encountered by incoming particles as they penetrate more deeply into the atmosphere.

The energy of the incoming particle beam determines how deeply into the atmosphere the beam will penetrate, and therefore at what altitudes the auroral light will be produced. The altitudinal distribution of the various atmospheric gases determines which kinds of gas the incoming particles are likely to strike and thereby cause to emit auroral light. The characteristics of the various gases (atoms and molecules) are crucial because those characteristics determine the colors of light emitted. Density also is important because in the high atmosphere, where the density is low, the processes that generate light are little affected by collisions between the atoms and molecules present. However, lower down, where the air is more dense, these collisions drastically modify the light-generating processes, enhancing some and quenching others.

Section 7.1 HOW ATOMS AND MOLECULES IN THE ATMOSPHERE PRODUCE AURORAL LIGHT

Plate 1 shows the radical difference between sunlight and auroral light. Sunlight has a continuous spectrum; every hue of each color in the rainbow is present (violet, indigo, blue, green, yellow, orange and red), and each blends into the next to form a multicolored but continuous array of light.[1] By contrast, auroral light lacks most hues, and its color comes from narrow regions in the light spectrum. The narrow regions, called lines and bands, are at specific places in the spectrum, and generally well separated from one another.

The easiest and most precise way to define the color of light is to specify its wavelength. The unit of wavelength used here is the angstrom unit (abbreviated A), equal to 1/100th million of a centimeter (10^{-8} cm or 1/10 nanometer). The term 'red light' describes light over a wavelength range from about 6000 A to nearly 7000 A, but the statement '6300 A red line' tells precisely what hue of red is involved. Notice in Plate 1 that the full visible spectrum extends from about 4000 to 7000 A, and that the light at the two ends is hard to see.

A major reason for describing the color of light by its wavelength is that light comes in quantized units called photons or light quanta, and each photon is of a specific wavelength. Specification of the wavelength of a photon tells not only what color it is but also the exact amount of energy it carries, and that is important because each photon is but a packet of energy—in the form we call light energy. When a photon strikes the eye it transfers energy to one of the eye's sensors, which

1. Not quite continuous, because a few thin, dark lines appear, owing to the absorption of light at specific wavelengths by solar gases.

then converts the energy to another form that can be carried through nerves to the brain, telling the brain that the photon did arrive.

How much energy does a photon carry? The answer comes from knowing the photon's wavelength because the wavelength and the energy are simply connected by an easily remembered relationship: the energy of a photon in electron volts is equal to 12,345 divided by the wavelength in angstroms.[2] Thus, for example, a red photon of wavelength 6300 (these photons produce the auroral red line) has an energy of 1.96 electron volts (eV), and a photon of ultraviolet light at wavelength 3000 A carries an energy of 4.12 eV. This example illustrates that light toward the red end of the visible spectrum carries the least energy, and that towards the violet end the most. (This is the reason that ultraviolet light is so much more damaging to human skin than other colors of light: the ultraviolet photons carry more energy that can disrupt the skin's cells.)

The significance of the facts that light comes in units of photons, and that each photon carries a set amount of energy, becomes more apparent as we now examine what happens when an incoming auroral primary strikes an oxygen atom in the high atmosphere. The light that results is an important part of the aurora we see.

Initially, the oxygen atom is in what is called its ground state, a condition of minimum energy. When an incoming auroral primary electron (or an energetic secondary electron) strikes the oxygen atom, it can hit so hard that the impact ejects an electron from the atom. The ejection requires an energy exchange of at least 13.6 eV. The atom then becomes a positively charged oxygen ion, and its internal structure is radically modified in the process. A less disruptive encounter can occur that also transfers energy from the incoming electron to the atom and modifies it slightly; that is, it leaves the atom whole but with altered internal arrangement. This is the process called excitation. In excitation, the atom accepts only certain amounts of energy. The rules of quantum mechanics come into play here, and they require that the oxygen atom must remain in its ground state or take in

2. Actually, the number is 12396, but 12345 is easier to remember and almost as accurate.

only a set amount of energy that will raise the atom to a new state. An oxygen atom can accept exactly 1.96 eV, the energy required to raise it to what is called its first 'excited' state. Alternatively, the atom can accept 4.17 eV, in so doing rising to its second excited state. Other allowed states exist at energy levels even higher, but elevation to those states is unlikely.

Oxygen atoms excited to states above the ground state are not happy oxygen atoms; like sleepy dogs suddenly jolted awake, they much prefer to go back to their original states. The dog may lay his head back down and again go to sleep, but the oxygen atom cannot return to the ground state without ejecting the energy it received. In possession of an extra 1.96 eV, the atom returns to the ground state by casting the excess energy out in the form of a light photon. In this overall process, the energy that began as kinetic energy carried by a primary electron moving into the atmosphere has transferred to a form of energy that binds atoms together, and thence to the light energy carried by a photon. The wavelength of the resulting photon is that dictated by the little 12,345 formula above, 6300 A. This is the wavelength of the oxygen red line in the aurora, and this one process—excitation of oxygen atoms to the first excited state, and the subsequent emission of photons carrying 1.96 eV—is the basic cause of the Type D all-red aurora, and the red upper coloration of the Type A aurora.

But there is more. The incoming electron also is allowed to transfer to an oxygen atom exactly enough energy to raise the atom to its second excited state, 4.17 eV. This atom is even more unhappy than the one elevated to the first excited state; it desperately wants to go back to the ground state. As shown in Figure 7.1, the atom has two possible photon-emitting routes to the ground state. The obvious path is direct: the emission of the full 4.17 eV in a single photon, one of wavelength 2972 A. However, the quantum mechanical rules that govern the life of atoms discourage this path, so the atom usually takes the other route. It falls to the first excited state by emitting a photon with exactly the energy difference between the first and second excited states, 2.21 eV. That photon has a wavelength of 5577 A. This yellowish-green emission is called the oxygen green line. It is the brightest single emission in the aurora

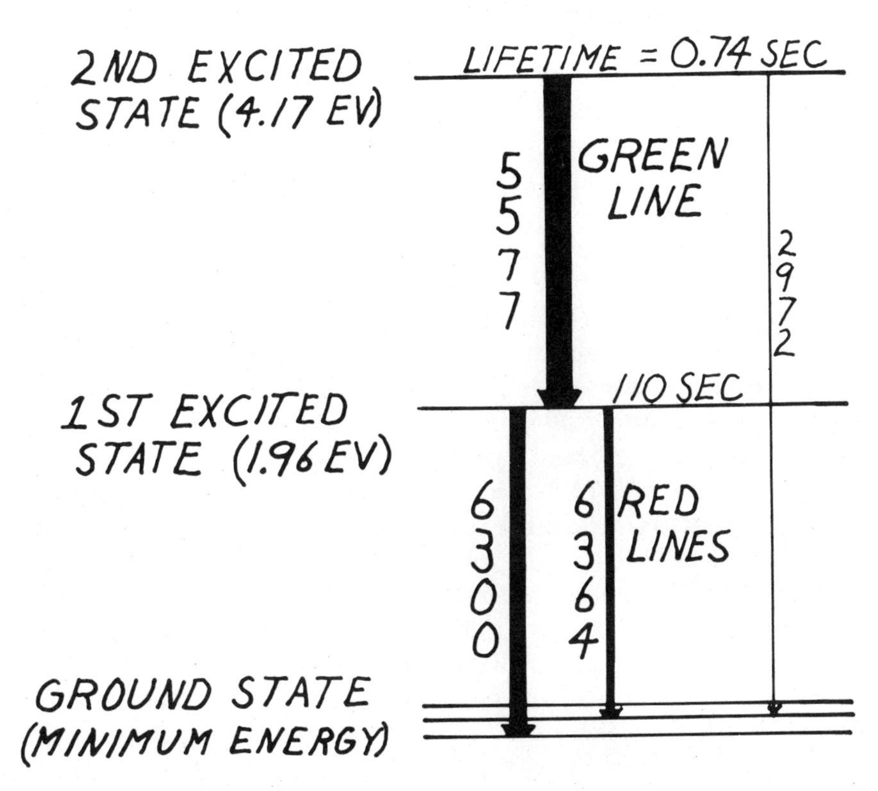

Figure 7.1 Auroral emission from the oxygen atom, giving rise to the 5577 A green line and the red lines at 6300 A and 6364 A.

most usually seen at the auroral zone, and the oxygen green line at 5577 A is responsible for virtually all the green light in the aurora. It produces the all-green Type C aurora and the green parts of Types B and E auroras. In fact, every aurora, even those that appear blood-red, contains green light produced by this transition of oxygen atoms from the second to the first excited states.

Having emitted the 5577 A photon in its fall to the first excited state, the oxygen atom can complete its trip to the ground state by emitting another photon of wavelength 6300 A. It would seem than that each time an oxygen atom emits a green photon at 5577 A it should also emit a red one at 6300 A, yet that does not always happen. The reason is that the oxygen atom can pause for a long time in the first excited state, and while resting there bounce into another atom or molecule. The encounter can carry off the excess 1.96 eV, thereby allowing the oxygen atom to fall to the ground state without emitting a red photon. If a collection of oxygen atoms is raised to the second excited state, half of the collection will emit green photons and fall to the first excited state in a time of 0.74 sec. The lifetime (or also the half-life) of this state is said to be 0.74 sec. However, the lifetime of the first excited state is 110 sec, and that is a very long time. Even so, if the collection of oxygen atoms in the first excited state happens to be located very high in the atmosphere (above 150 km), the density there is low enough that the atoms are likely to emit 6300 A photons and in so doing create red auroral light. If the collection is lower down, at an altitude less than 150 km, the atmosphere is dense enough that oxygen atoms in the first excited state usually will collide with other constituents before their 110-sec lifetime is up. No 6300 A photons are emitted, and the red line emission is said to be quenched.

A minor complication of little consequence is evident in Figure 7.1. The diagram shows that oxygen atoms in the first excited state can fall to the ground state by emitting a photon of either wavelength 6300 A or one of 6364 A. That is possible because the ground state actually consists of three sublevels, each with slightly different energy. Transitions from the first excited state to two of the three levels are allowable. The most probable is the transition that emits 6300 A photons, and approximately three of them appear for each 6364 A photon. Since the two emissions occur in a set ratio, most discussions ignore the fact that the red emission from oxygen atoms actually is composed of two red lines rather than one. Everybody is supposed to understand that the term 'oxygen red line' or 'oxygen 6300 A emission' includes both 6330 A and 6364 A photons.

The above discussion deals only with the oxygen atom, yet the ideas expressed therein extend to all other atoms and molecules present in the atmosphere that contribute to creating auroral light. The choice of the oxygen atom for this discussion is an obvious one because oxygen atoms do produce a great part of the light we see in the aurora. The oxygen atom is the most abundant species of gas in the high atmosphere, and it produces the brightest single line emission in the aurora, the 5577 A green line. Furthermore, this line is at a wavelength where the human eye is most sensitive. By contrast, the oxygen red line at 6300 A falls in a part of the spectrum where the eye is only one-fifth as sensitive. Other strong emissions in the aurora also occur where the eye responds poorly, in the blue and violet portion.

Many of the other auroral emissions come from molecules rather than atoms, and these emissions are somewhat more complex. A hint of the reason for the added complexity lies in the foregoing discussion of the ground state of the oxygen atom, a state composed of three levels of almost the same energy. Two of the three are involved when an oxygen atom emits the two line emissions lying near one another at 6300 and 6364 A, in the red part of the spectrum. When the possible energy states of molecules are examined, they are seen to be composed of many closely spaced levels. When these multiple levels are involved in the production of light, the result is emission of many photons possessing slightly different wavelengths. The photons of one wavelength produce what is called a line emission, and an overall collection of related and closely spaced line emissions is called a band emission. The individual line emissions that make up a band may be so close together that they appear to overlap, hence the name band is appropriate. Several strong band emissions occur in the aurora. One of the strongest is a band near 3914 A that comes from ionized nitrogen molecules. This band emission is approximately as strong as the oxygen 5577 A green line, but it occurs in the violet part of the spectrum where the human eye lacks sensitivity. Another somewhat weaker band produced by ionized nitrogen molecules lies near 4278 A in the blue-violet region. It too provides little recognizable auroral light. Neutral nitrogen molecules produce several band emissions in the red part of the spectrum, at wavelengths somewhat longer than that of the oxygen red line (see Plate 1). These emissions are primarily responsible for the Type B and Type E auroras, those with red lower borders.

Section 7.2
ALTITUDE EFFECTS

One of the key factors that determines the color of the aurora is the distribution of the various species of gas contained in the atmosphere. Near the earth's surface the atmosphere has a uniform composition of permanent gases, primarily molecular nitrogen (N_2), 78 percent, and molecular oxygen (O_2), 21 percent. Making up the remaining 1 percent, in order of decreasing concentration, are carbon dioxide (CO_2), argon, neon, helium, krypton, xenon, hydrogen, methane (CH_4), and nitrous oxide (N_2O). This distribution is uniform up to an altitude of about 100 km. At higher levels, the energetic photons contained in ultraviolet sunlight break up molecules into atoms, and also the different species tend to sort themselves out by gravity, the lighter ones rising to the top. In the very highest part of the atmosphere, above approximately 500 km, the light hydrogen and helium atoms predominate. Below that altitude, down to about altitude 200 km, oxygen atoms are the most prevalent. Nitrogen molecules well outnumber other species in the height range 100 to 200 km, and oxygen atoms and molecules make up most of the other gas present. Oxygen and nitrogen molecules far outnumber all other species in the height range 60 km to 100 km. Just from knowing this distribution, a person can correctly guess that in the lowest aurora, that occurring between 60km and 100 km, most of the light should come from oxygen and nitrogen molecules. Higher up, in the region 100 km to 200 km, nitrogen molecules and oxygen atoms should produce most of the auroral light. Still higher yet, above 200 km, most of the light should come from oxygen atoms, and the somewhat lesser number of nitrogen molecules there should also contribute. In the very highest part of the atmosphere, the high relative abundance of hydrogen and helium should also lead to auroral emissions. However, these are weak and have little if any effect on the apparent color of aurora.

The density of the atmosphere within the height range where aurora occurs also is a critical factor in determining the color of auroras. At the earth's surface, the number of molecules in a cubic centimeter of air is huge: about 2.6×10^{19} (26 billion billion). The density decreases with altitude so that at 50 km above the surface the number of gas particles is 1000 times less, and at 100 km the number is 2 million times less than at the surface. At altitude 200 km, the number of gas particles per unit volume is a billion times less than at the earth's surface. We tend to think of the whole region above altitude 100 km as being an excellent vacuum, but at altitude 200 km the number of gas particles per cubic centimeter still is about 10 billion. (Out in the solar wind, the density truly is low, only about 5 particles per cubic centimeter.)

Although many gas particles are present at altitudes of 150 km and above, the number is small enough that collisions between particles are relatively infrequent. Such collisions affect the colors of the aurora because they can remove atoms and molecules from excited states by extracting the energy that otherwise would go into the emission of photons. Because of the long lifetime of the oxygen atom in its first excited state, 110 sec, this atom is especially susceptible to collisions that remove its extra energy. At altitudes above 200 km, collisions are rare enough to have little effect, but at lower altitudes the collisions quench the 6300 A red line very effectively. An oxygen atom located in the lower parts of the aurora simply is unable to emit the 6300 red line. Frequent collisions in the lower parts of the aurora have many other effects as well, because they allow transfer of energy from one molecule or atom to another and thereby create many complexities in the production of auroral light.

Section 7.3 THE CAUSES OF THE VARIOUS AURORAL COLOR TYPES

Rather than to examine the color types in alphabetical order, a more logical approach is to start at the top of the atmosphere and work down, in essence following along with the incoming electrons and protons. Those that lack sufficient energy to penetrate the atmosphere deeply create the Type D all-red aurora.

Type D, all-red aurora: When an incoming particle stream is composed of electrons with energies of but about 500 eV or protons of energy of but a few keV, the stream can penetrate only to an altitude of about 200 km. The incoming particles strike the atoms that are present, hydrogen, helium and oxygen. The impacts on hydrogen and helium do not create enough visible light to have much effect on the color of the aurora produced. Mainly, the incoming particles produce red light by transferring enough energy to oxygen atoms to raise them to the first excited state, and far fewer of them to the second excited state. The oxygen atoms then emit some green light at 5577 A, but much more red light at 6300 A. The result is aurora that appears red to the human eye, Type D.

Type C, green aurora: An incoming particle stream composed of electrons with energy 10 keV or protons of energy of several hundred keV can penetrate to an altitude of approximately 100 km. The incoming beam will lose most of its energy toward the end of its path, in the height range 100 km to 150 km. Here the incoming electrons and protons strike mostly nitrogen molecules and oxygen atoms. The impacts on the nitrogen molecules ionize some of the molecules and also raise the resulting ionized nitrogen molecules to excited levels. Instantly, the ionized nitrogen molecules emit strong band emissions in the blue and violet part of the spectrum. However, these emissions are in a region of the spectrum where the human eye responds poorly. The excited neutral nitrogen molecules also yield band emissions in the red, but these, too, are in a part of the spectrum where the eye lacks sensitivity.

The aurora created in this height range looks green because of excitation of oxygen atoms. As is the case higher up in the atmosphere, the incoming electrons and protons are more likely to raise oxygen atoms to the first excited state, and fewer of them to the second excited state. Those raised to the second state remain there, on the average, for 0.74 sec before emitting green photons of wavelength 5577. They then are in the first excited state, ready to emit red photons of wavelength 6300 A. However, the long lifetime of this state, 110 sec, usually allows the oxygen atom to bump into another atom or molecule before emitting a photon. Thus, the oxygen red line is quenched in this altitude range, and the numerous oxygen atoms emit only green light. The 5577 A emission is very strong in this height range, and it falls right near the peak of the eye's sensitivity. The result is the Type C all-green aurora.

Type A, green aurora shading to red at the top: Type A aurora is a sort of mixture of Types C and D, and it is caused by the same processes. If the incoming electron or proton beams are composed partly of particles with enough energy to reach down to altitude 100 km and partly of particles of lesser energy, then they produce the green aurora of Type C in the lower region and the red aurora of Type D higher up. The degree of redness at the top of Type A auroras will of course depend on the relative proportions of energetic particles and less energetic particles in the incoming beams, and the incoming electrons produce far more visible light than the protons.

Types B and E, green auroras with red lower borders: The Type B and E auroras[3] are caused by incoming particle beams with sufficient energy to penetrate the atmosphere below altitude 100 km, perhaps in the extreme to as low as 70 km. In this lower region, oxygen atoms are very few, and the air is composed mostly of nitrogen and oxygen molecules, in the ratio of about 4 to 1. The incoming beam generates the Type C green aurora above 100 km, and below that altitude excites the nitrogen and oxygen molecules so that they produce band emissions, mostly in the red part of the spectrum. The nitrogen molecules in particular emit strongly in the red, creating an easily seen red lower border. These red emissions are instantaneous, and for that reason, the red tips of the fast-moving Type E auroras often appear to move ahead of the green upper parts created by the comparatively slow emission of the green 5577 A photons—on the average, 0.7 sec later. If two of the Type B or E auroras lie close together, an observer is likely to see the spectacular sight of red-tipped rays moving in opposing directions, eastward along one form and westward along its partner.

Type F, blue-purple sunlit aurora: Sometimes observers see very tall auroral rays that have a blue-purple cast. These typically occur during winter morning twilights, or during early fall and late spring, the times when sunlight is falling on the high atmosphere above the observer. The blue-purple coloration of the Type F aurora

3. Recall that these are fundamentally identical: Type E is Type B that moves rapidly.

is the result of an enhancement of the emission from ionized nitrogen molecules when they are exposed to sunlight. It is partly a two-step process. First, the incoming electrons or protons strike neutral nitrogen molecules and ionize them. Some of the ionized molecules also are excited to energy states that produce band emissions in the violet and blue (especially the 3914 A and 4278 A bands). Second, the ionized nitrogen molecules absorb energy from the sunlight at precisely the same wavelengths and immediately re-emit the energy. This process is called resonance scattering. Because of the resonance scattering of sunlight by ionized nitrogen molecules, they emit more light than they normally would were sunlight absent. Of all the constituents in the high atmosphere, only ionized nitrogen molecules have the capability to emit substantial visible light through the resonance scattering process.

Section 8
THE AURORA SEEN FROM SPACE

The advent of satellites in 1958 opened up a new era in auroral investigation. An immediate effect was renewed interest in the aurora because space scientists recognized that this phenomenon could give insight into processes occurring in the new region to be explored: the near-earth space environment. Satellites also provided a new tool for investigating auroras. They could, for example, fly over auroras (and through the highest of them) carrying instruments that would measure the characteristics of the incoming particle streams that generate the aurora. Manned and unmanned satellites also provided platforms that allowed visual observation and photography of the aurora from above.

Section 8.1
PHOTOGRAPHS FROM SATELLITES

Astronaut Owen Garriot took the first actual photograph of an aurora from space on September 5, 1973, using a Hasselblad camera, hand-held but braced against a window of the Skylab. Shown here in Plates 4, 5, 11 and 12 are other recent photographs of southern hemisphere auroras acquired on the NASA shuttle mission STS-39, flown in April and May 1991. Viewed from slightly above and well equatorward, the auroras in the photographs look much as a ground observer would expect. They hint at the field-aligned character of the discrete aurora, and the way the lower borders of the bands hug the earth's surface brings home the fact that the forms maintain a constant altitude as they follow along the earth's curvature. The images show the red upper portion typical of Type A aurora, a characteristic more noticeable in space photographs than in those taken from the ground.

Section 8.2 AURORAL IMAGERY USING SCANNING PHOTOMETERS

Much information about the aurora has come from optical scanners aboard several polar-orbiting satellites, most notably from a series of vehicles launched in the early 1970s by the United States Air Force for the purpose of acquiring images of cloud cover both at night and during the daytime. The images obtained by these satellites present a truly global view of the aurora that provides a perspective totally different from that obtainable from the earth itself.

Sec. 8.2.1 Acquisition of DMSP Auroral Images

The satellites flown by the air force's Defense Meteorological Satellite Program (DMSP) have been placed in nominally circular orbits 850 km above the earth, high above the aurora. Each has been in a sun-synchronous orbit; that is, each orbit is in a plane with fixed orientation relative to the earth-sun line. Some of the satellites fly along the noon-midnight meridian; others traverse along the dawn-dusk meridian. The period of each orbit is 102 min.

Each DMSP satellite carries a scanner device capable of detecting light in the wavelength region 4000 to 11,000 angstroms. Thus the detectors are sensitive to light seen by the human eye and also to some infrared light. A rotating mirror is arranged so that it allows the detector to accept a narrow beam of light from an ever-changing direction at right angles to the satellite's path. As the mirror rotates, the device scans the region below the satellite, a swath 3000 km (2000 mi) wide.

The output of the scanning photometer can be assembled into a photograph-like image just as the scan lines on a television set produce an image on the TV screen, or as a typist assembles a page of text, character by character, one line at a time. But only one 'photograph' of the aurora in one hemisphere is produced on each satellite pass, and that image takes about 20 minutes to acquire and assemble. Thus, each resulting image is not a snapshot. It is more like the image produced by an old-time pan camera, in that one end of the picture is taken well after the other. Just as small boys used to run from one end of a group photo to the other while the camera panned, thereby getting recorded twice, the aurora can change radically during the required 20 minutes to yield an appearance that departs from reality.

The combination of the satellite's forward motion, the rate of mirror rotation, and angular width of the scanning beam permits a maximum resolution of 3 km directly below the satellite. This resolution is insufficient to record small-scale characteristics of auroral forms, but adequate for detailing major features. The greatest limitation of the DMSP images is that they are available only every 102 minutes, so profound auroral changes can occur during the interim. Nevertheless, the fact that each image contains such a large fraction of the auroral oval makes the DMSP imagery both interesting and valuable. The quality of the DMSP images is excellent and the view they provide exciting (at least to scientists in the appropriate field).

Among the first of the DMSP auroral images published are the ones shown in composite form in Figure 8.1. The satellite that obtained these images was in the noon-midnight sun-synchronous orbit. It acquired the right-hand image first, then as the earth rotated eastward beneath the orbit, the three others in order. During this sequence most of North America was enjoying clear, moonless sky, but during the last two passes moonlight was bright enough to show the continental outlines of Alaska and eastern Siberia, as well as cloudiness in the North Pacific Ocean.

A narrow region of aurora to the north of Montreal appears in the first image, while the second one shows a very broad extent of aurora, illustrating a remarkable variation in the display during the course of less than two hours. Note that the north-south extent of the aurora seen in the second image is greater than the distance between Canada and Mexico.

Sec. 8.2.2 Views of the Auroral Oval and of Auroral Substorm Variations

Through intensive study of auroral photographs acquired

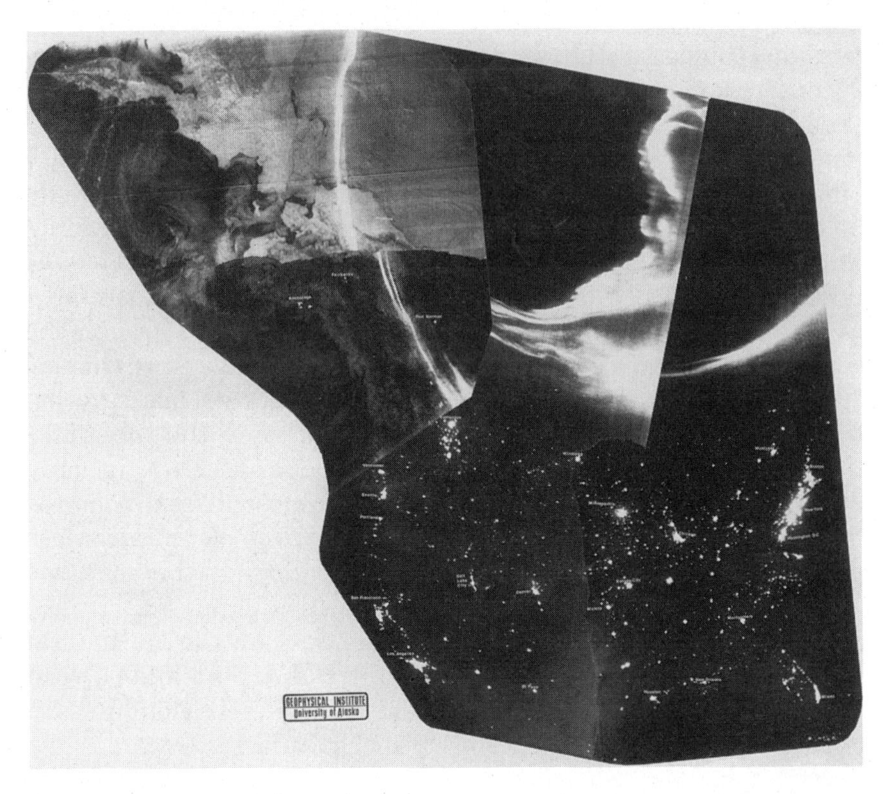

Figure 8.1 A montage of four DMSP images acquired by a satellite in polar orbit, showing the aurora at times 102 minutes apart and all of the United States except for Hawaii.

by ground-based all-sky cameras, auroral scientists previously deduced the existence of the auroral oval and the auroral substorm. When the DMSP auroral images became available, they confirmed the existence of the oval and the substorm. Furthermore, interpretation of the images requires these two concepts.

Figure 8.2 illustrates the power of the DMSP imagery. This image shows at left the lights of Siberian cities connected by the Trans-Siberian Railway, a spectacular sight in itself. Above that is a circle indicating the area of aurora that a ground-based observer can see. The satellite image presents a much broader view. This image shows a large portion of the auroral oval during the peak of a major substorm, and also the three main types of auroral forms that appear within it. To the north of Siberia appears the faint hydrogen arc, a highly uniform and slightly striated fea-

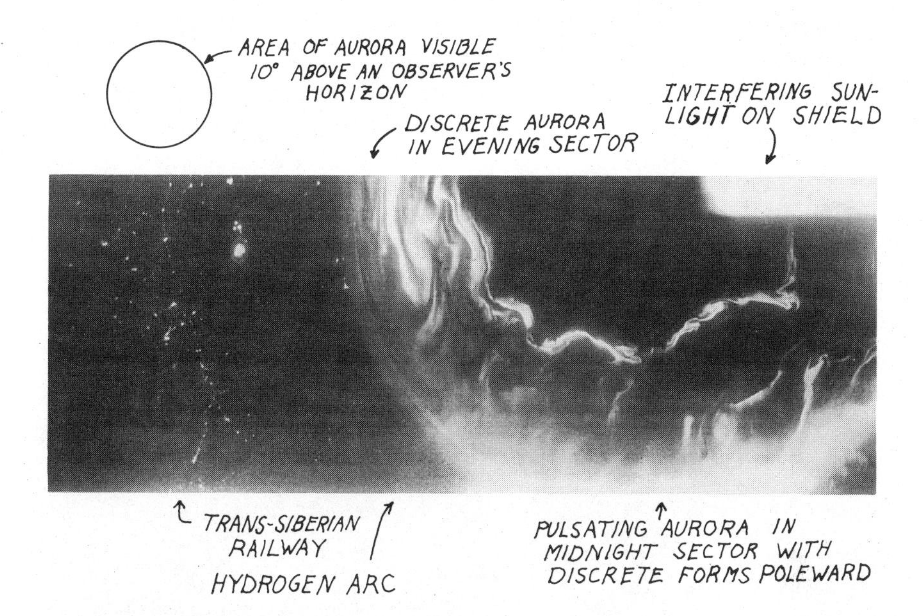

Figure 8.2 DMSP image of a very large auroral display. Note the circle at upper left which illustrates the area of aurora easily seen by an observer on the ground, that portion more than 10° above his horizon.

ture. North of that is a broad region of discrete aurora. At lower right, representing the midnight and early morning portion of the display, is an extensive region of pulsating aurora, and to the north of that a contorted zone of discrete aurora. By itself, this one image presents a summary overview of much of the material presented earlier in Sections 5 and 6.7.

Sec. 8.2.3 A Short Catalog of DMSP Images

The DMSP images presented here are approximately in order of increasing levels of substorm activity. The direction to the sun is toward the top of each image. A white band at top or at upper right is due to sunlight glinting off a light shield. Note the changes in the diameter and width of the oval as the activity increases.

Figure 8.3 Pass Number 0449, 10 December 1972—The image shows a narrow auroral oval with substorm activity just commencing on the midnight meridian. The lights at lower left are those of cities in Great Britain, and others toward lower center are cities in Scandinavia.

Reflected light from the sun shield creates the white blur at lower right.

Figure 8.4: Pass 0831, 1 June 1973—Image of a somewhat broadened oval showing a narrow band of discrete auroral forms in the evening sector (at left) and pulsating aurora in the midnight and morning sector (at bottom and at right). A small region of discrete aurora also appears poleward of the main oval in the evening sector.

Figure 8.5: Pass 1094, 25 January 1973—Acquired near the

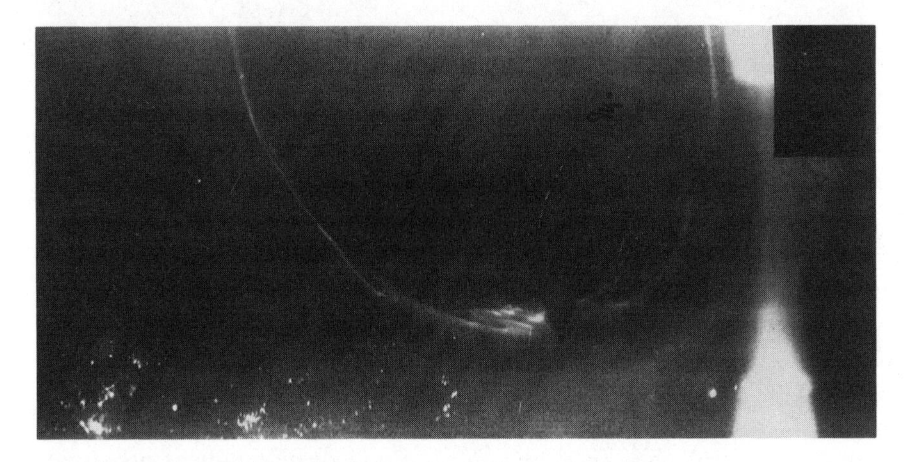

Figure 8.3 DMSP Pass 0449 on December 10, 1972, showing a thin auroral oval just as minor substorm activity commences at the midnight meridian.

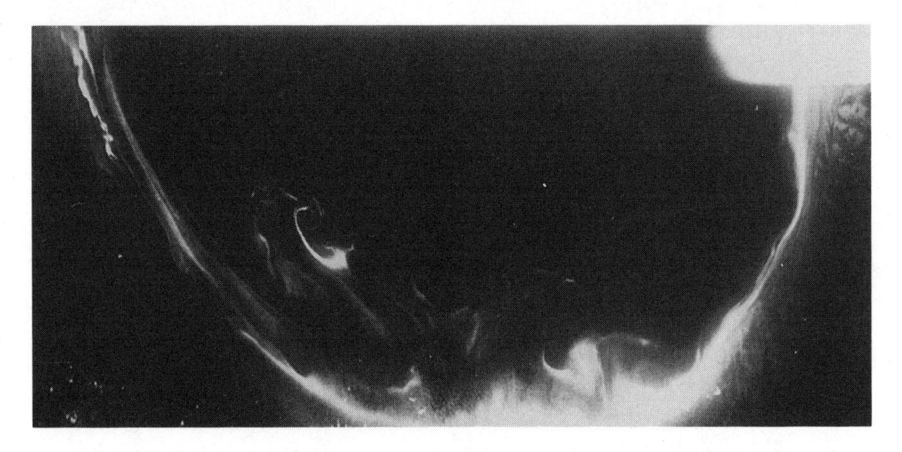

Figure 8.4 DMSP Pass 0831 on June 1, 1973, during a period of moderate substorm activity.

peak of the expansive phase of a moderately large substorm, this image features extensive discrete aurora in the evening sector, including an array of arcs folded back on itself and, to the north of that, a spiral form. The discrete aurora extends over into the midnight meridian, lying poleward of a broad region of pulsating aurora that extends southward beyond the image. The white vertical line is an instrumental artifact.

Figure 8.6: Pass 911, 12 January 1973—A complex spiral-like discrete aurora appears at upper left, in the evening sector, and discrete aurora extends poleward of it and across to the north of an extensive array of pulsating aurora in the near-midnight sector. As sometimes occurs during high levels of activity, fingers of the pulsating portion of the display extend poleward beyond the main part of the oval. This image was acquired near the end of the expansive phase of a very large substorm.

Figure 8.7: Pass 1096, 24 January 1973—The image shows an extensive display of pulsating aurora in the midnight and morning sectors. The vertical streaking at lower right is an instrumental artifact.

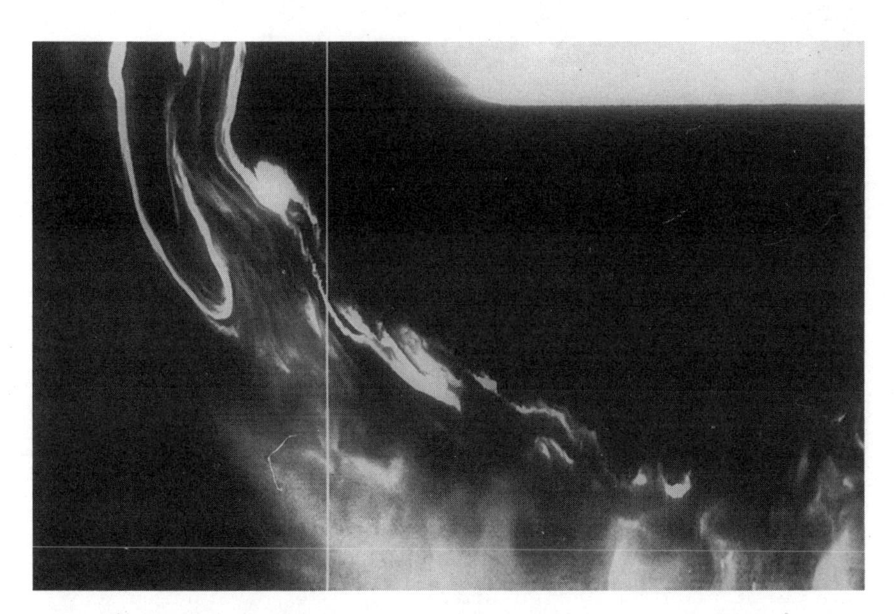

Figure 8.5 DMSP Pass 1094 on January 25, 1973, during high activity. It shows extensive discrete aurora in the evening sector and extending eastward past midnight, with pulsating aurora in the midnight and morning sectors (at bottom and lower right).

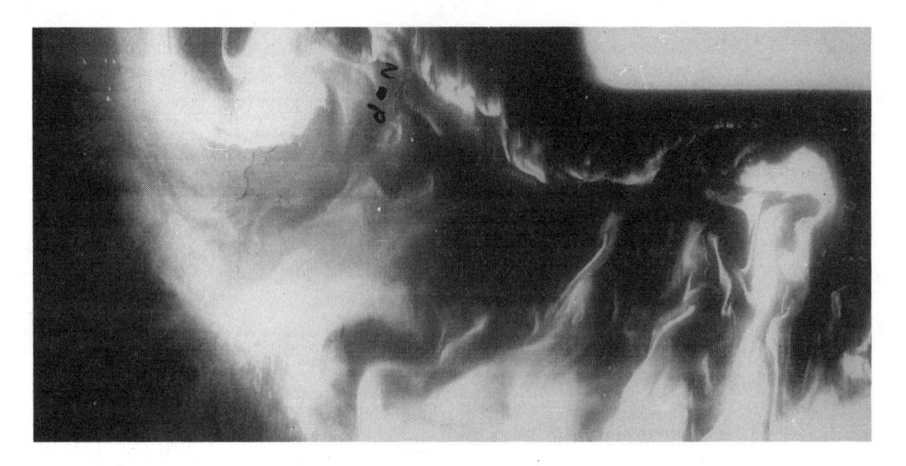

Figure 8.6 DMSP Pass 911 on January 12, 1973, acquired during partial moonlight so that continental featues show below an extensive aurora; pulsating portions (sometimes referred to as 'torches') extend poleward out of the main portion of the oval.

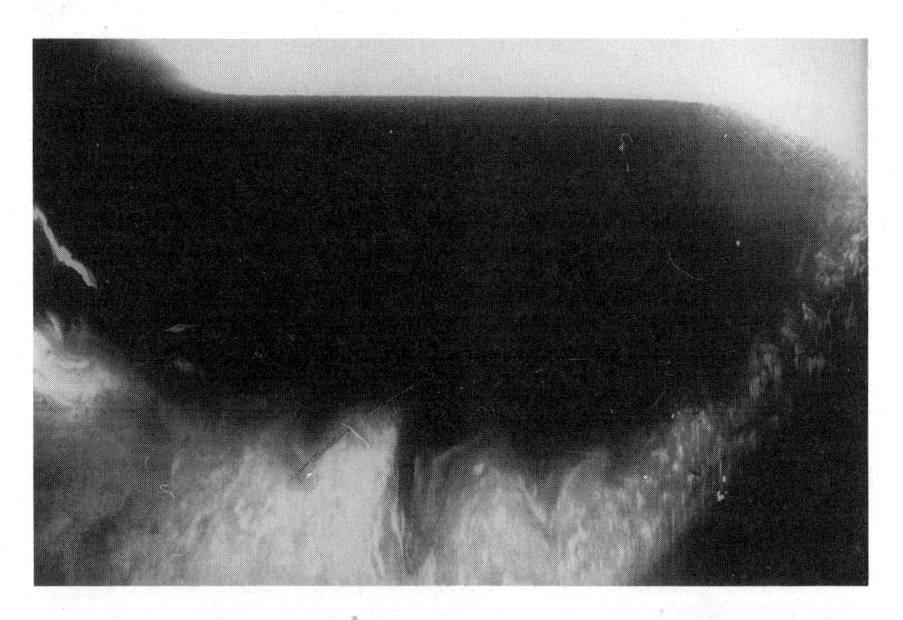

Figure 8.7 DMSP Pass on January 24, 1973, showing a broad expanse of pulsating aurora in the midnight and morning sector and at far left, a small discrete aurora in the evening sector.

Section 9

CONTROL OF THE AURORA BY THE EARTH'S MAGNETIC FIELD

The earth's magnetic field (also called the geomagnetic field) not only controls where aurora occurs, it is essential for its production. Only those planets that have their own magnetic fields and atmospheres, planets such as Jupiter, Saturn, Uranus and Earth, also have auroras. Essential also to the production of earthly auroras are electric fields, mostly of transient nature.

Human beings in their natural state—before the discovery of the lodestone, and before they began to use electricity in their daily lives—had only one force field to worry about: gravity. By the time a human child learned to walk, he or she knew all necessary about that simple field, the fact that it points down everywhere. Electric and magnetic fields are more complex; they point in various directions, but for reasons that can be understood if pondered. They also cause charged particles to move in directions not expected on the basis of experience gained from simply living in a gravity field, or from playing billiards.

The relevant essence of the matter for the student of aurora is that the earth's magnetic field is essentially that of a bar magnet (also that of a current loop or a magnetized sphere), and that charged particles flow unimpeded in the direction of a magnetic field but hesitate to move great distances in a direction perpendicular to the magnetic field. Therefore the bar magnet-shape of the earth's magnetic field causes the charged particles to enter the

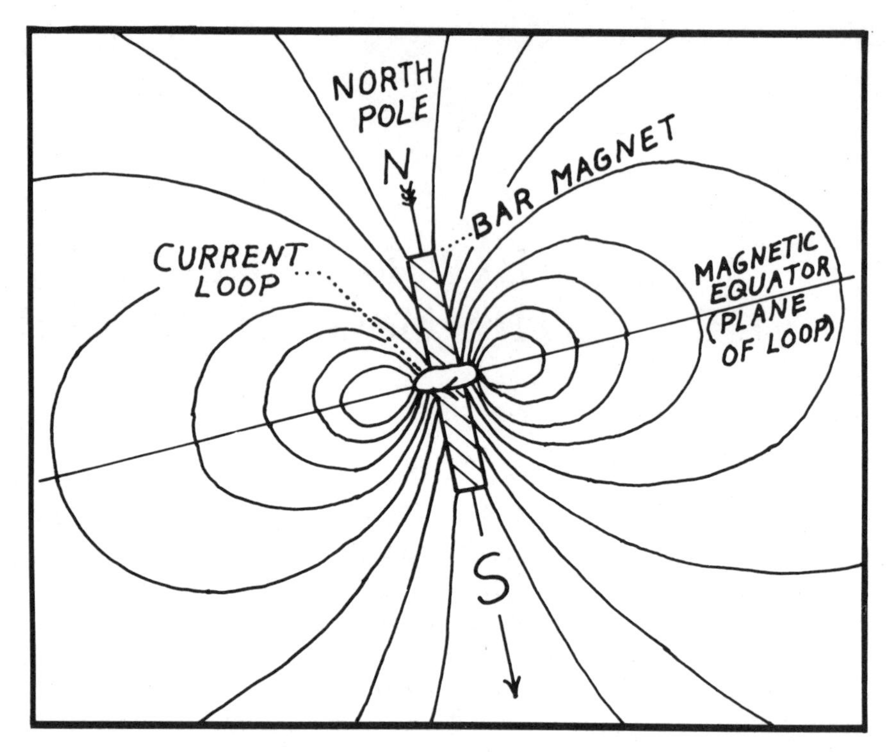

Figure 9.1 The magnetic field of a bar magnet and a current loop; the field's configuration also can be seen by sprinkling iron filings near a bar magnet, or a spherical one, as in Figure 1.1.

atmosphere at high latitudes, and it prevents them from entering near the equator. The incoming auroral particles receive their primary guidance from the geomagnetic field, but electric fields influence them also. The electric fields act in concert with the geomagnetic field to alter the directions in which charged particles move, and the electric fields accelerate (speed up or slow down) the particles.

Section 9.1 THE CAUSE AND SHAPE OF THE EARTH'S MAGNETIC FIELD

Why is the earth's magnetic field like that of a bar magnet? And why is the magnetic field of a bar magnet like that of a current loop? Figure 9.1 illustrates the configuration, one so important to the aurora that an answer to these two questions is in order. We start with the fundamental fact that a line current has a magnetic field at right angles to the direction of the current, as shown in Figure 9.2. The magnetic field at any point in space is the sum of all the magnetic fields at that point, regardless of their source. A consequence of the additive characteristic of magnetic fields is that if a line current is bent around to form a circular loop, the magnetic field near the loop takes on the shape shown in Figure 9.1. Figure 9.1 is a two-dimensional representation of a three-dimensional entity (the magnetic field) that is perfectly symmetrical about the axis of the loop—assuming the loop to be a perfect circle. If the thumb of the right hand points in the direction of the current, the fingers curl in the direction of the magnetic field—the right-hand rule so familiar to physics students.

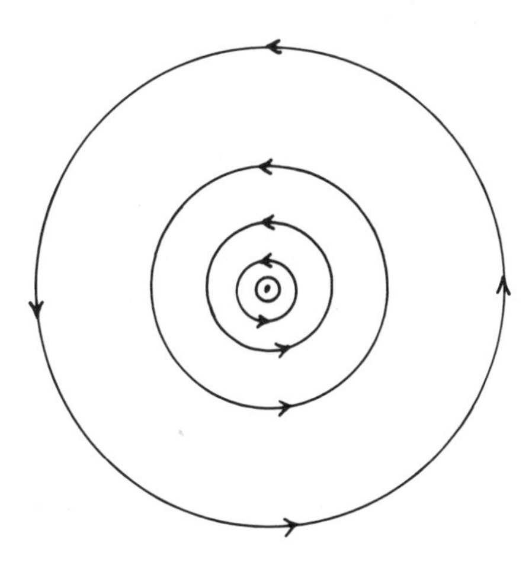

Figure 9.2 The magnetic field of a line current directed out of the page. The spacing between the lines is shown inversely proportional to the strength of the field, that is, the closer the lines, the stronger the field.

Now consider a bar of steel. It is composed of many iron atoms, and the electrons attached to the nucleus of each atom are travelling in circles about the atomic nuclei. Because each electron is charged, its circular motion constitutes a current loop that generates a tiny magnetic field of the shape shown in Figure 9.1. In an unmagnetized bar of steel, the electron circuits are in random directions, and so the individual magnetic fields cancel each other. Thus, the bar has no net magnetic field. But when the bar becomes magnetized, the process involves alignment of the atoms so that the axes of a large portion of the individual current loops point in the same direction, as illustrated in Figure 9.3. The individual magnetic fields now add to one another, and the result is a magnetic field that, well outside the steel bar, is identical to that of a simple current loop. The earth's magnetic field at a point well outside the earth is the same, for the reason that

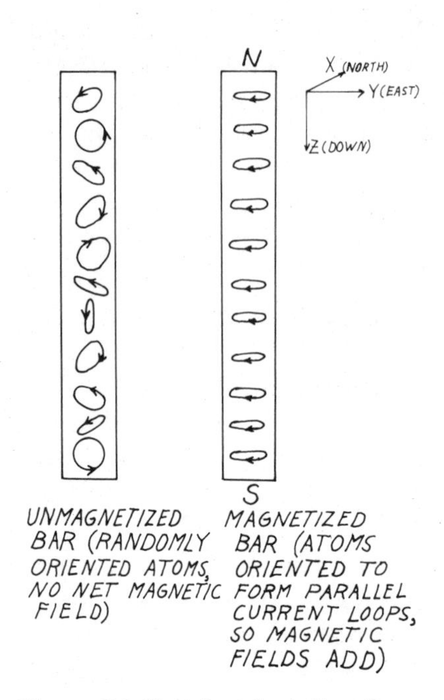

Figure 9.3 Sketches depicting the random orientation of atoms in an unmagnetized iron bar (left), and, at right, in a magnetized bar.

this field is created by current loops deep inside the earth, mostly in planes approximately parallel to the equator.

If a bar magnet of the appropriate strength were placed near (but not exactly at) the center of the earth such that its axis were on a line through Thule, Greenland, and Vostok, Antarctica, the magnet's magnetic field would closely approximate that of the earth. The points of intersection with the earth's surface are called the geomagnetic poles (also the dipole poles). These poles are more than just mental constructs because, at distances of several thousand kilometers above the earth's surface, the geomagnetic field does indeed look like that of a bar magnet aligned as described, and it is symmetrical about a line drawn through these poles. This field determines the location of the northern and southern auroral zones, and it also is the framework that anchors the northern and southern auroral ovals to the earth system. The northern auroral oval maintains an offset symmetry about the north geomagnetic pole, and the southern auroral oval does the same about the southern geomagnetic pole. Thus, the geomagnetic field is the main overriding control that determines the general regions where aurora will appear.

The geomagnetic field is almost that of a current loop, and a significant characteristic of the magnetic field of a current loop is that its strength falls with the cube of the distance from the center of the loop. That is, if the field strength is B at one distance from the loop's center, the field strength twice as far away will be only an eighth as powerful (since B/(2 x 2 x 2) = B/8). But the combined magnetic field of two or more mutually displaced current loops has a strength that falls with the fifth, seventh, and even higher powers of the distance, depending on the relative configurations of the loops. All this is relevant to the earth's permanent magnetic field because that field is generated by complex circulation of hot, ionized (and therefore electrically charged) material in the earth's core. Since the greatest portion of the circulation is roughly parallel to the earth's equator, the most significant part of the geomagnetic field is in the form of a single current loop. This component of the geomagnetic field therefore has a strength that falls away with the cube of the distance from the center of the earth. However, not all the circulation in the core is parallel to the equator. Some of it is in cells lying above and below the equator and oriented north-south or in other directions. The contribution to the geomagnetic field by these cells falls away with distance from the center of the earth according to the fifth, seventh and higher powers of distance. The consequence is that, at the earth's surface, the observed geomagnetic field is more complicated than that of a simple current loop or bar magnet. Yet higher up, only a few thousand kilometers above the surface, the geomagnetic field is almost strictly that of a bar magnet. That field up there is what the aurora's causative incoming charged particles 'see' as they approach the earth.

For the reasons given above, the earth's magnetic poles—the places where a balanced compass needle points vertically—do not correspond to the geomagnetic poles. Indeed, the actual field observed over the earth's surface is sufficiently complex that a person who wishes to navigate by compass needs a magnetic map that displays the direction a compass needle will point at each location. That direction is determined by the orientations and sizes of the circulation loops in the earth's core, and also in some places by the proximity of iron ore bodies or other magnetized rocks lying near the earth's surface. Some ore bodies are strongly enough magnetized to cause radical departure of a compass needle from its expected direction, in some cases even strong enough to reverse its direction.

To make matters worse for the navigator, the circulation of currents within the earth's core is not constant with time. The main features of the irregular magnetic field drift slowly westward so that every few years it becomes necessary to update magnetic maps. Over the course of centuries, substantial changes have occurred: the magnetic poles have drifted around, the declination (angle of the magnetic field from the true north direction) has shifted, and the locations of the auroral zones probably have changed. During past millions of years, the earth's magnetic field even has undergone reversals; that is, the north and south magnetic poles have changed hemispheres. These reversals evidently are caused by extreme changes in the circulation currents within the earth's core.

The region outside the earth and its ionosphere where the geomagnetic field is strong enough to control the motion of charged particles is called the *magnetosphere*. Within the inner magnetosphere—the region within a few earth radii of the surface—the geomagnetic field has a shape very much like that of a current loop or bar magnet, and here the field is strong enough to exert a major control on the motion of charged particles. Farther out, in the outer magnetosphere, the control is still strong but not in complete domination. At the fringe of the magnetosphere, at what is called the *magnetospheric boundary* or the *magnetopause,* a tug of war occurs between the geomagnetic field and the charged particles. Here, the magnetosphere gives way to the solar wind, and the shape of the magnetospheric boundary and its location depends on the intensity of the wind. The wind interacts with the magnetosphere at the magnetopause to distort the boundary and sweep it back in the anti-solar direction, as shown in Figure 9.4. In this diagram the earth appears as a small sphere located near the left of the drawing. Roughly 20 earth radii in diameter, the magnetosphere extends approximately 10 earth radii toward the sun, and tails back hundreds of earth radii in the direction away from the sun, just like the tail of a comet.

Figure 9.4 implies that the magnetosphere is a complex object, as indeed it is, with several identifiable regions named by those who have investigated this vast portion of near-earth space by means of instruments carried aboard satellites. Most important for the purpose here is the general configuration of the geomagnetic field in the magnetosphere—like that of a bar magnet near the earth, but distorted farther out and stretched back in the anti-solar direction. Most of the charged

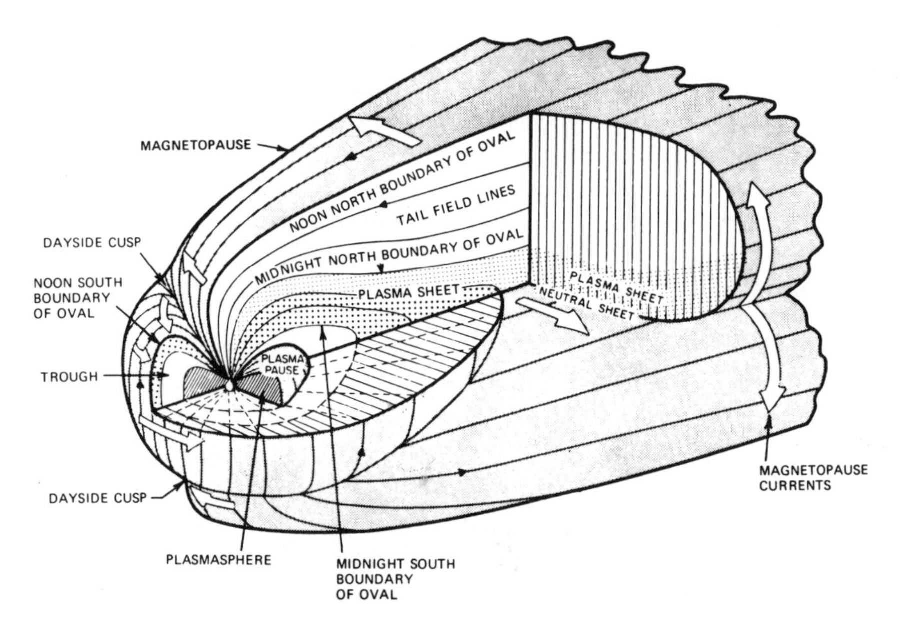

Figure 9.4 A 3-dimensional, cut-away view of the earth's magnetosphere drawn by Walter Heikkila, showing various regions that have been recognized and named. The earth is the small white circle toward the left.

particles that enter the atmosphere to generate auroras probably are transiting through this highly distorted outer portion of the magnetosphere that lies behind the earth in the direction away from the sun.

Section 9.2 GUIDANCE OF CHARGED PARTICLES BY THE GEOMAGNETIC FIELD

The behavior of charged particles in electric and magnetic fields is more complex than that of particles or objects acted on by only mechanical forces. A mechanical force, such as gravity or the force exerted on a pool ball by a cue stick, causes motion in the direction in which the force is applied—a rock released from the hand falls to the ground because the force of gravity is downward toward the center of the earth, and the pool ball always rolls in the direction the cue stick hits it (at least if the pool player is adept enough to hit the center of the ball).

By contrast, a charged particle immersed in a magnetic field receives influence from the field only when the particle moves. That is, the magnetic field exerts no force on the particle as long as that particle is at rest. However, when the particle is moving, the force exerted is at right angles both to the direction of the particle's motion and to the direction of the magnetic field, as illustrated in the top portion of Figure 9.5. This odd state of affairs has two consequences. If a moving charged particle is placed in a magnetic field oriented in the same direction as the particle's motion, then the magnetic field exerts no force on the particle.

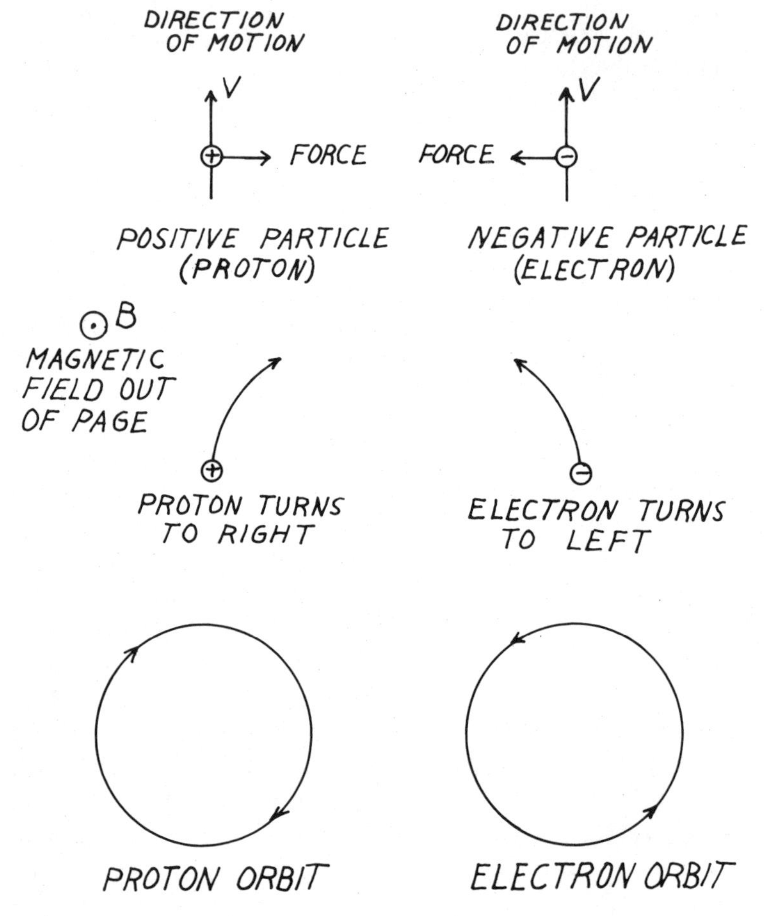

Figure 9.5 Top: The force (F) on a charged particle in a magnetic field (B), shown directed out of the page) is at right angles both to its direction of motion (V) and the direction of the field. The force's direction depends on the sign of the particle's charge. The sketches at center show how the force creates a curving path that results in circular motion, as shown at the bottom.

The particle keeps on moving as though the magnetic field were not there. However, if the initial motion of the particle is at right angles to the direction of the magnetic field, the force exerted, since it is at right angles to both the magnetic field and the particle's motion, causes the particle to execute a circle, as indicated in the bottom portion of Figure 9.5.

Unless the initial direction of motion of a charged particle is in one of these two extremes—either exactly parallel to the direction of the magnetic field as shown in Part A of Figure 9.6, or exactly perpendicular to the direction of the magnetic field as shown in Part B—the particle will move along a helical path. The axis of the executed helix is along the direction of the magnetic field, as is shown in Part C of the diagram. The helix's diameter depends on the particle's mass and speed, on the electrical charge it carries, and on the strength of the magnetic field.

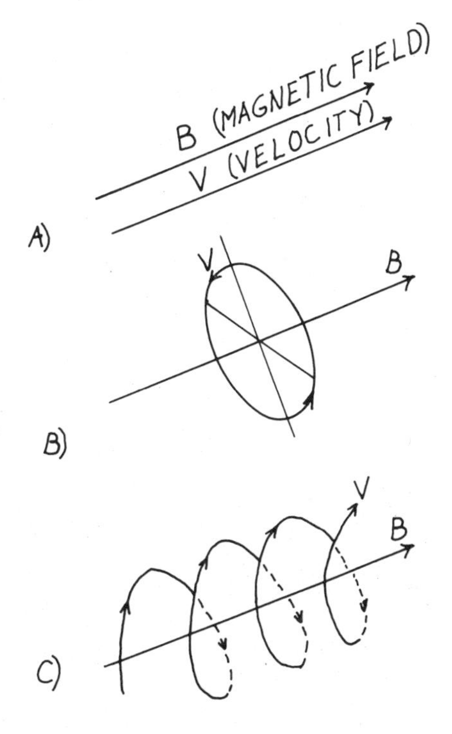

Figure 9.6 Part A: A charged particle moving in a direction V that is parallel to the direction of the magnetic field B is not affected by the field and continues to move in that direction. Part B: A charged particle moving exactly perpendicular to the direction of the magnetic field B travels in a circle. Part C: A charged particle moving neither parallel to nor perpendicular to the direction of the magnetic field executes a helix with axis parallel to B, the direction of the magnetic field.

Herein lies the key to understanding why the geomagnetic field guides the charged particles that generate the aurora into the polar atmospheres. Once these particles leave the solar wind and enter the region where the geomagnetic field pertains (the magnetosphere), they travel in helical paths along the direction of the geomagnetic field, and the magnetic field's shape brings them to the atmosphere at high latitudes. The geomagnetic field's constraint

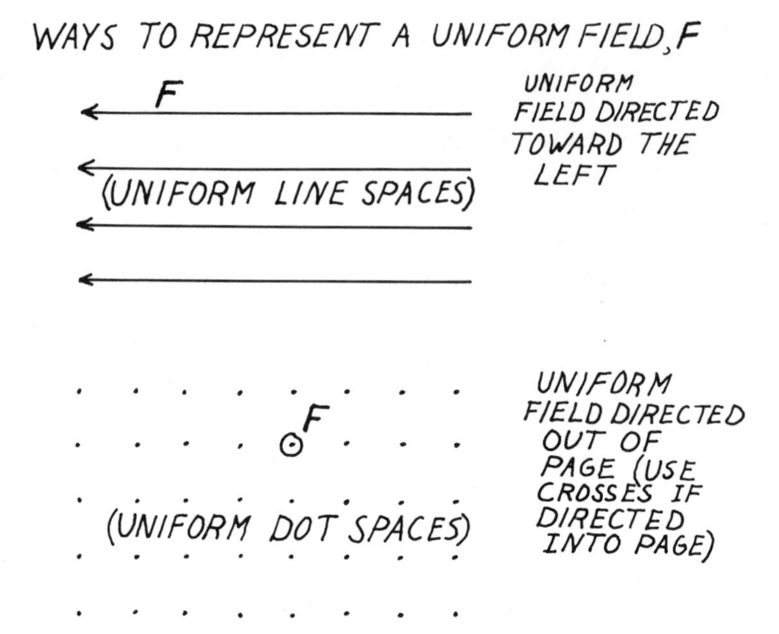

Figure 9.7 Two representations of a uniform magnetic field, one directed toward the left at top, and one directed out of the page at bottom. Note the even spacing of the lines at top and of the dots at bottom.

on the motion of the particles as they enter the atmosphere also is the reason why discrete auroral forms are aligned along the direction of the local geomagnetic field, nearly vertical at the auroral zones.

In thinking about and in describing magnetic or other kinds of force fields, a common practice is to visualize "field lines" or draw them on paper, a general technique already employed here in discussion and diagrams such as those in Figure 9.1 and Figures 9.2 to 9.6. If done properly, the lines drawn on paper represent both the direction and strength of a field; for example, see Figure 9.1. Lines drawn with uniform spacing, as in Figure 9.7, represent a uniform magnetic field—one that does not vary in any direction. Compare this diagram to Figure 9.1 which shows by the spacing of the lines that the magnetic field weakens with distance from its source, in this case a line current.

This general procedure makes use of the mental construct of 'magnetic field lines.' Magnetic field lines do not physically exist—they are only a product of the human mind—but a person watching rayed aurora easily forgets that fact when he sees the aurora actually marking out the 'field lines.' Everyone is supposed to understand that the field lines are not real physical entities and, with that understanding, we glibly say that charged particles travel along magnetic field lines and that they travel in helical paths aligned along the field lines, very much like beads on a wire. The analogy is quite precise because the diameter of the helices is usually quite small compared to the other dimensions being considered. For example, the electrons that come into the atmosphere to create an auroral ray execute helical paths with diameters of only about 1 meter, whereas the diameter of the ray created by the entire particle stream is 1 to several kilometers and the length of the visible ray may exceed 100 km.

The essence of charged particle motion in the earth's magnetosphere is as described above, and that "bead on a wire" concept is the main thing to remember. But real life is not quite so simple, because the description given above pertains only to a *uniform* magnetic field with no other force fields present, fields such as the earth's gravity field or any electric fields that may be lurking about.

The geomagnetic field is not a uniform field: its field lines converge as they approach the earth's surface, and that means that the field is stronger there than farther out. This gradient in field strength causes charged particles to drift away from the field lines they would like to follow. Because the diameter of the circular path a particle follows is smaller in a strong field than in a weaker one, a particle out in the magnetosphere will drift sideways across the field lines in the fashion illustrated in Figure 9.8. As they move along the geomagnetic field toward the earth, incoming particles that carry negative charge (such as electrons) drift slowly eastward, and protons or other positively charged particles drift slowly westward. This fact is at least part of the reason why auroras tend to be aligned east-west as well as along the direction of the geomagnetic field. The incoming particle streams tend to spread themselves out toward the east and west.

Although the incoming charged particles are very light, they do experience a scant pull by the earth's gravity field that

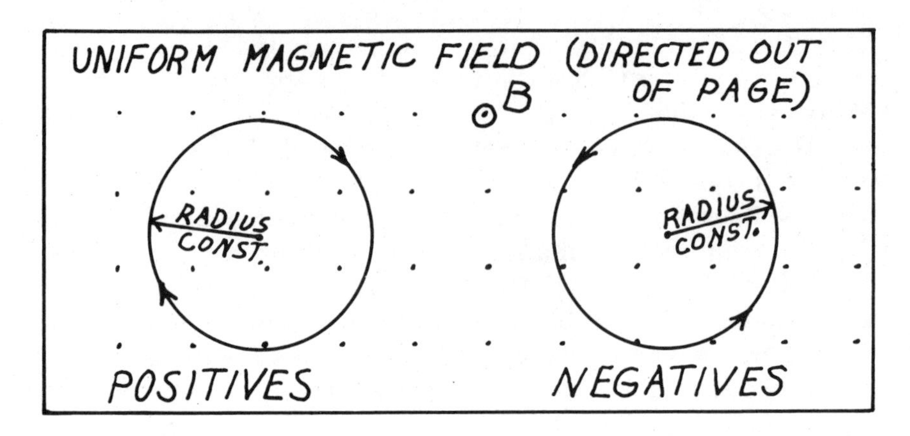

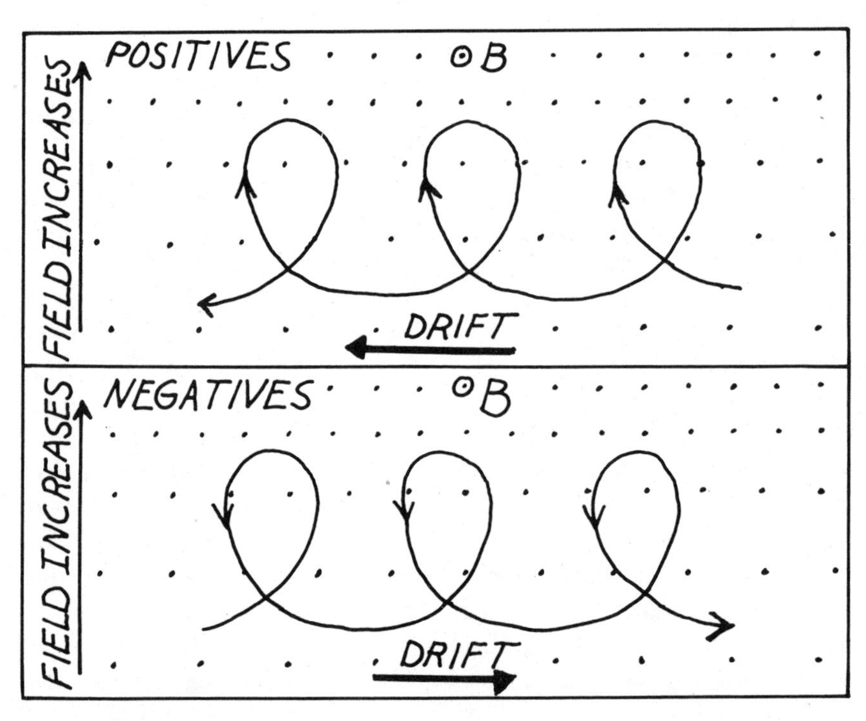

Figure 9.8 Top: The direction of the circular motion of positively and negatively charged particles in a uniform magnetic field (directed out of the page). Bottom: The drift of the particles in a nonuniform field, one increasing in strength toward the top of the page. The particles drift in opposite directions.

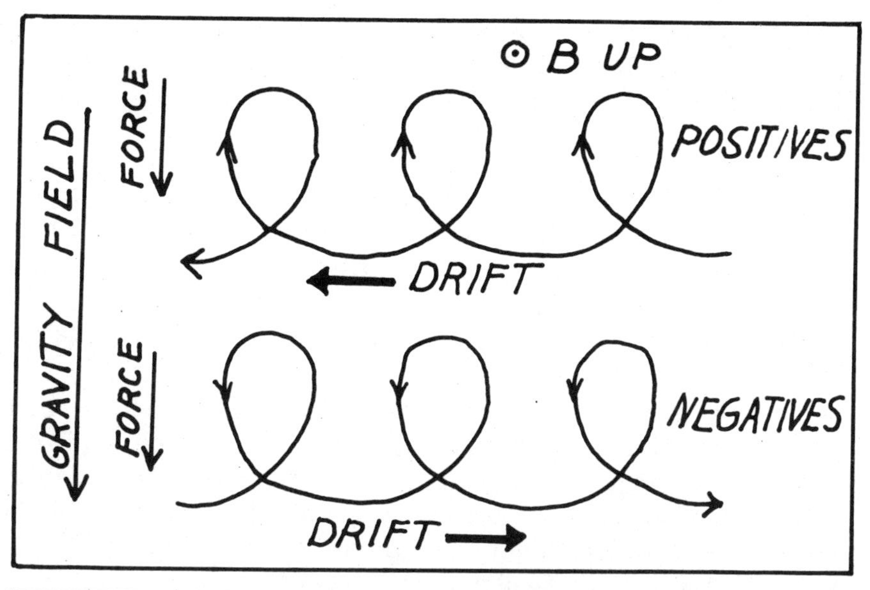

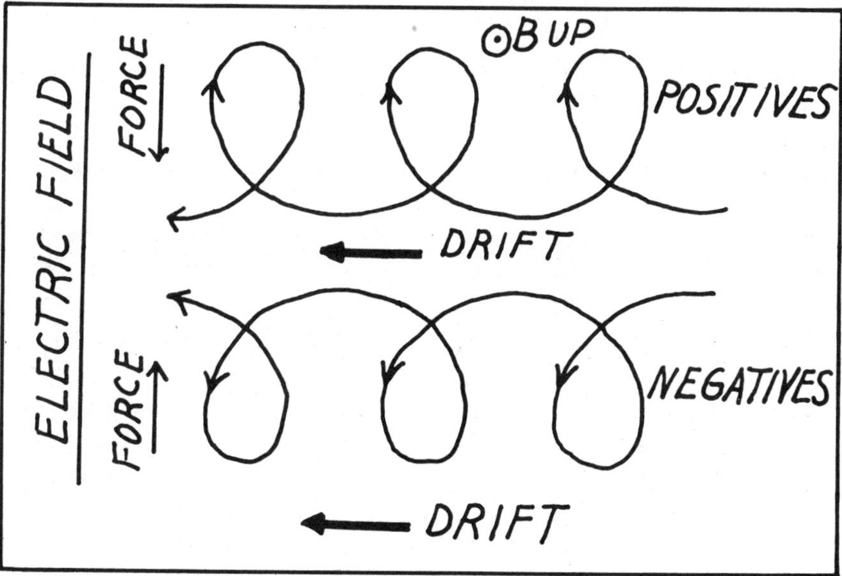

Figure 9.9 Top: Oppositely charges particles drift in opposite directions in the presence of a magnetic field and a field (such as gravity) that exerts force independent of the sign of charge on the particle. Bottom: The drift of charged particles in an electric field perpendicular to a magnetic field is in the same direction, at right angles to both fields. (The magnetic field B is shown directed up out of the page in both panels.)

slightly alters their circular paths about magnetic field lines. The result is an additional minor drift of positively charged particles to the west and negatively charged particles toward the east, as shown in the top part of Figure 9.9.[1] (When looking at this diagram and others similar, think of the illustration as representing the situation in the earth's equatorial plane; there the geomagnetic field points northward. If you then 'look' from the equatorial plane in the outer magnetosphere down along the geomagnetic field lines to the north polar region, note that east and west are preserved, and that the direction away from the earth maps to north, while the direction inward toward the earth maps to south.)

If electric fields are present—and such fields indeed are present in the magnetosphere—they too can cause a drifting of charged particles across magnetic field lines. The part of an electric field directed parallel to the magnetic field lines will affect charged particles, either speeding them up or slowing them down, depending on the sign of charge carried by the particles. If the electric field is partly oriented perpendicular to the magnetic field lines, that perpendicular part does not affect the speed of the particles but it does cause them to drift across the magnetic field lines, and in a direction at right angles to the direction of this part of the electric field; see the lower part of Figure 9.9. Both negative and positive particles drift in the same direction, typically to the east or west because the main electric field in the outer magnetosphere is mostly directed either inward toward the earth or outward.

In summary, charged particles in the magnetosphere try very hard to follow along magnetic field lines like beads strung on wires. In the process the guidance provided by the magnetic field neither speeds up nor slows down the particles in the long run; that is, the total energy carried by the particles does not change. The particles also undergo drifting motions, mostly toward the east or west because the geomagnetic field is not uniform (it is stronger in some places than in others), and because of the presence of the earth's gravity field and large-

1. A similar drift, positive particles westward and negative particles eastward, occurs because of the curvature of the geomagnetic field. Called the curvature drift, it is actually more important than the gravity drift.

scale or small-scale electric fields. These drifts perpendicular to the magnetic field neither add or detract from the energy the particles carry; in the vernacular of the physicist, the motion is unaccelerated. Acceleration occurs only in the presence of an electric field that has a component directed parallel to the magnetic field lines.

Whereas the geomagnetic field in the magnetosphere has a grand overall structure—we can think of it as a single all-pervasive entity—the electric fields are more complex. Electric fields derive from distributions of electrical charge, each charge having its own electric field, as is shown in the top portion of Figure 9.10. The electric field of a positive charge (carried by a proton) points outward; the field of a negative charge (carried by an electron) points inward. Charged particles immersed only in an electric field become accelerated and move as shown in the bottom portion of Figure 9.10: positively charged particles in the direction of the electric field, and negatively charged particles in the opposite direction. The acceleration of auroral primaries in the magnetosphere is by such a cause, one that may involve relatively stable electric fields or highly transient ones.

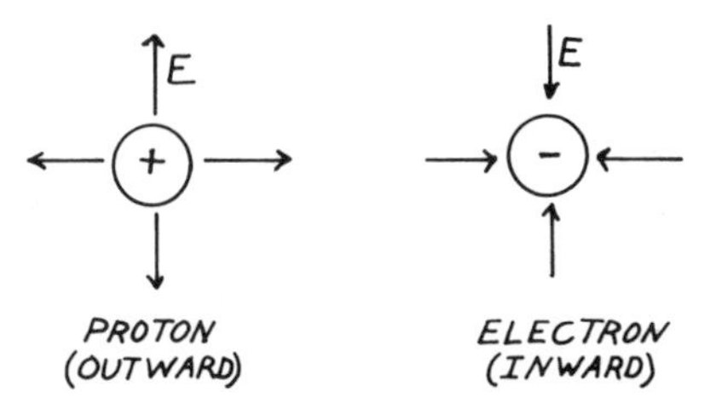

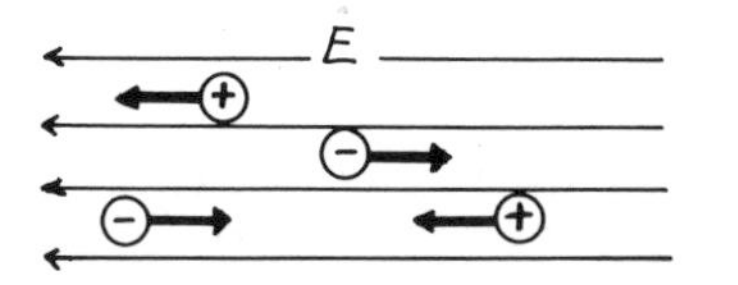

Figure 9.10 Top: Electrical charges give rise to electric fields. The field is directed outward away from a positive charge, and inward toward a negative one. Bottom: In the absence of a magnetic field, positively charged particles move in the direction of an electric field (E), and negatively charged particles move in the opposite direction.

Section 9.3 THE GIANT MAGNETIC BOTTLE UP IN THE SKY

Highly relevant to particle motion in the magnetosphere is the curious concept of the magnetic bottle, a magnetic field configuration that locks charged particles inside. A magnetic bottle consists of a magnetic field arranged so that its magnetic field lines converge at both ends, such as shown in Figure 9.11. It

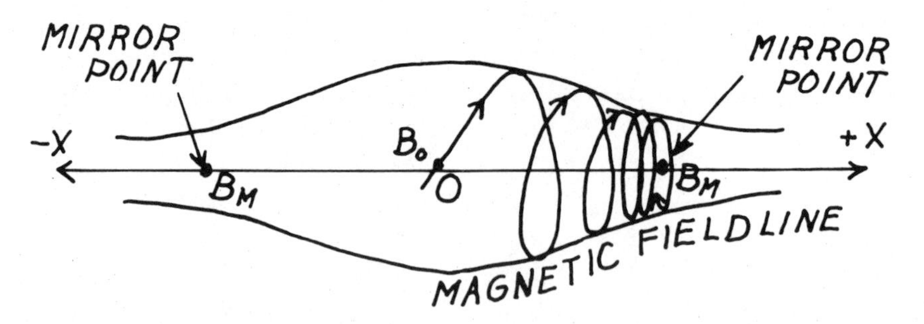

Figure 9.11 The configuration of a magnetic bottle. Charged particles spiral along until they reflect from the mirror points (B_M), and thereby remain confined. The magnetic field at each mirror point is higher than elsewhere inside the bottle.

is able to confine a charged particle because the particle's motion in a helical pattern is governed by a rule that requires kinetic energy (energy of motion) be conserved, regardless of what direction the particle is moving.

A charged particle moving in a magnetic field has a certain kinetic energy that depends only on the particle's mass and the square of its speed. If the particle is moving through a converging magnetic field, it gyrates in increasingly tighter circles, but the speed of its circular motion increases. The kinetic energy involved in that circular motion therefore also increases. But the total kinetic energy must remain constant, and so the kinetic energy of circular motion must come at the expense of the particle's forward motion along the field lines. Reduced forward kinetic energy requires reduced forward speed, so therefore the particle's forward motion slows down. The farther the particle moves toward one end of a magnetic bottle, the slower it goes in that direction. The particle finally undergoes one very fast last circular gyration, its kinetic energy all locked up in circular motion. That last gyration does not take place in a perfectly uniform magnetic field, and because of that fact, the motion during part of the last orbit has a slight component in the direction opposite to that in which the particle was previously travelling. The particle continues to spin, but now it begins to move in the reverse direction, slowly at first, but ever faster and faster because it is moving in the direction of a diverging magnetic field. The particle is said to have *mirrored* from the end of the magnetic bottle, at a point called the *mirror point*. Just where the mirror point occurs depends both on the strength and configuration of the magnetic bottle and the kinetic energy carried by the charged particle and its direction of motion when first injected into the magnetic bottle. After reflection from its mirror point, the particle proceeds to the other end of the bottle, and the process repeats. Since no energy is interchanged between the particle and its containing bottle, the particle will move back and forth forever—unless something else happens that alters the bottle or affects the particle's kinetic energy. This is a remarkable process, to be sure: a particle with mass that repeatedly changes direction without losing or gaining energy. In a world governed only by mechanical forces such a thing could never happen, but the rules of electromagnetism are different.

Through the proper arrangement of magnets, it is possible to build in the laboratory a magnetic bottle like the one shown in Figure 9.11. And if a bottle of that shape is bent so that its two ends are swung around toward each other, the resulting configuration is that of a cross-section of the geomagnetic field. Thus, sitting up there in the magnetosphere is a true magnetic bottle, one shaped like a doughnut placed around a marble-sized earth, and quite capable of trapping charged particles within it.

When charged particles enter the magnetosphere from the solar wind, some of them move down the geomagnetic field to the polar atmospheres as shown in Figure 9.12, and others penetrate deeper into the magnetosphere to get trapped in the magnetic bottle located there, commonly called the Van Allen belt after its discoverer, James Van Allen.

Section 9.4
ENTRY OF ENERGY AND CHARGED PARTICLES INTO THE MAGNETOSPHERE

The sun boils off electrons, protons (hydrogen nuclei), and lesser amounts of positively charged nuclei of elements heavier than hydrogen. These particles make up the solar wind, an outward flow that travels at speeds of 300 to 1000 kilometers per second and brings the particles to the earth several days after they are ejected from the sun. Also carried along in the solar wind is an irregular but somewhat

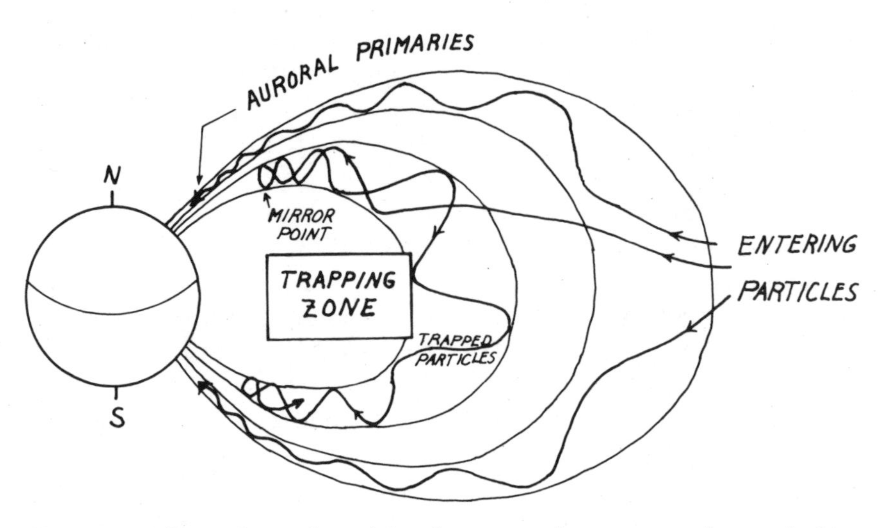

Figure 9.12 Some charged particles that enter the magnetosphere spiral down the magnetic field lines to the auroral zones where they strike the atmosphere and generate auroras. Others penetrate more deeply into the magnetosphere and enter its trapping zone where they remain for relatively long periods, bouncing back and forth between mirror points as they drift around the earth.

structured extension of the sun's magnetic field. The charged particles in the solar wind act as a sort of glue that attaches the weak solar magnetic field to the wind (the field is said to be 'frozen in') and carries it along. Thus the solar wind actually consists of a stream of charged particles in which a weak magnetic field is immersed and is therefore transported along with the particle stream.

As the solar wind approaches the earth, at a distance of approximately 30,000 km (a distance more easily visualized if expressed as 10 earth radii (R_e)) the charged particles in the wind begin to interact with the earth's magnetic field. On the sun-earth line, the geomagnetic field lines are approximately perpendicular to the flow of the solar wind, so the geomagnetic field impedes the wind. The oncoming particles try to penetrate the geomagnetic field but it turns them aside, positively charged particles to the east, negatively charged particles to the west. This induced east-west motion creates an electrical current in a direction that causes its magnetic field to add to the geomagnetic field. It is as though the pressure of the solar wind were compressing the geomagnetic field, making it stronger. The result is a sharp demarcation called the magnetosphere boundary. Inside the boundary lies the magnetosphere, a region where the geomagnetic field pertains and from which the solar wind is generally excluded. Outside the boundary, the solar wind continues to flow, unabated but diverted so as to move on around the sides of the magnetosphere.

On the sides of the magnetosphere, right at the boundary, the solar wind particles tend to attach to the outermost geomagnetic field lines much as though the field lines were weak, highly stretchy rubber bands. The flow of the solar wind along the magnetosphere boundary thereby drags the geomagnetic field lines back in the anti-solar direction. Indeed, the geomagnetic field lines do act like rubber bands, thick near the earth, but thinning outward until, near the magnetosphere boundary, they are so weak that they stretch easily. The result is that the solar wind sweeps the outer fringes of the geomagnetic field backward a great distance behind the earth. The drag of the solar wind also creates the visible tail on a comet, and the magnetosphere takes on a similar shape. The stretching is so extreme that the magnetosphere is aptly described as looking like a broom handle that has a pea-sized earth lodged within its sunward end. The tail of the magnetosphere extends far out in the anti-solar direction, well beyond the orbit of the moon.

Some energy and particles enter the magnetosphere through the dayside cusps. These are high-latitude regions where geomagnetic field lines extend out to meet the magnetosphere boundary at nearly right angles, as shown in Figure 9.4. But scientists believe that most particles and energy cross the magnetosphere boundary through the tail of the magnetosphere. A critical factor affecting the entry is the direction of the weak solar magnetic field carried within the solar wind. Some scientists believe that the direction possessed by the solar magnetic field when it arrives near the magnetosphere is determined by the direction the field had when it left the visible surface of the sun. Whatever that direction may be, and whatever its cause, the directionality usually persists for several days at a time. If the direction is generally southerly rather than generally northerly, then the situation portrayed in Figure 9.13 can pertain. Right at the magnetosphere's boundary, geomagnetic field lines attach to the field lines in the solar wind so that a magnetic connection develops across the boundary. The connection then provides a pathway for particles in the solar wind to penetrate the magnetosphere boundary and enter an unstable region located well behind the earth. What happens there is not known, but one view is that the kinetic energy carried by the particles becomes temporarily stored as magnetic energy which then releases suddenly during substorms and causes the acceleration of auroral primaries.

Although the details are not understood, it is clear that the extent to which solar wind magnetic field lines connect to those of the outer magnetosphere is critically important. That merging best occurs when the solar magnetic field is southward pointing, and the largest magnetic storms and the greatest

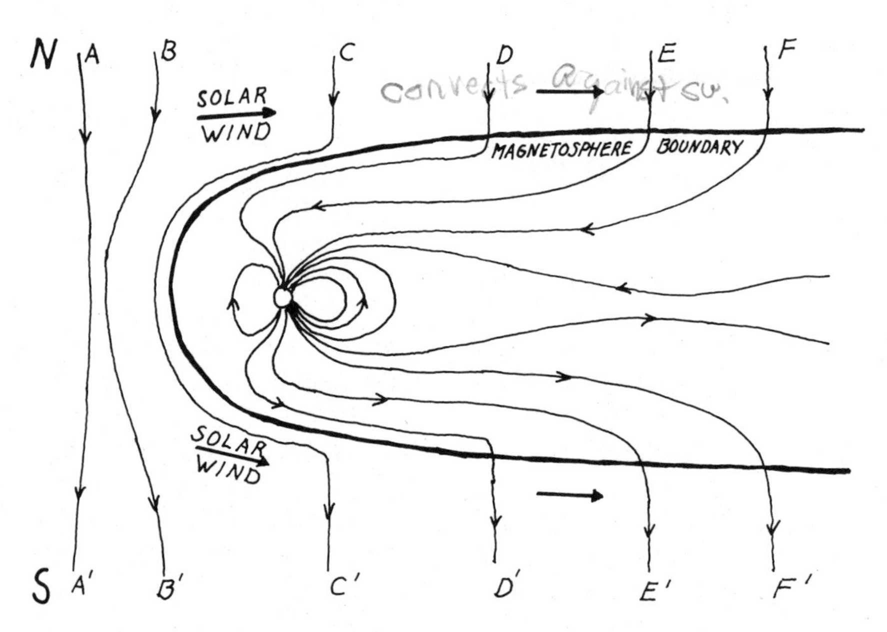

Figure 9.13 If pointed north to south, solar magnetic field lines carried in the solar wind can connect across the boundary of the magnetosphere to those inside, providing paths for incoming particles. As they sweep past the earth the field lines successively occupy positions listed in the sketch diagram as AA' to FF'.

auroral displays occur only when the solar magnetic field points nearly directly south—and also when the solar wind is moving most rapidly, and when the strength of the solar magnetic field at the magnetosphere boundary is at its strongest. Major auroral displays and major magnetic storms apparently never occur unless the solar magnetic field at the magnetospheric boundary is pointed primarily south.

By the resulting merging process, the magnetic field lines in the solar wind are connected across the magnetosphere boundary to some of the geomagnetic field lines within the magnetosphere. The solar wind particles now blow across the merged field lines. This flow generates electromotive force, just as does a conducting coil moving in a magnetic field—the principle by which an ordinary generator operates. Called a magnetohydrodynamic dynamo, the solar wind-magnetosphere generator operates on the same general principle, but diffuse flows of charged particles rather than wires form the electrical conductors that move through a magnetic field to generate the electromotive force. (Recall that 'electromotive force' is the term used to describe the force or electric pressure that causes a current to flow in a circuit. An ordinary automotive battery has an electromotive force of 12 volts, so we call it a "12-volt battery.") The solar wind-magnetosphere electrical generator has a variable electromotive force, weak at the times we think of as having low magnetic activity, and very strong during magnetic storms. During the greatest of storms the electromotive force may be as much as 100,000 volts.

The solar wind-magnetosphere generator establishes a voltage across the tail of the magnetosphere that, coupled with the geomagnetic field, drives a circulation of charged particles contained in the outer magnetosphere in two cell-like convection patterns, as illustrated in Figure 9.14. This convective motion in the equatorial plane of the outer magnetosphere actually involves both particles and magnetic field lines to which they are connected, and so the outer portion of the magnetosphere circulates with the particle motion. This motion of the outer magnetosphere projects right on down along the geomagnetic field lines to the earth's polar regions, as is depicted by Figure 9.15. One of the effects at the auroral zone

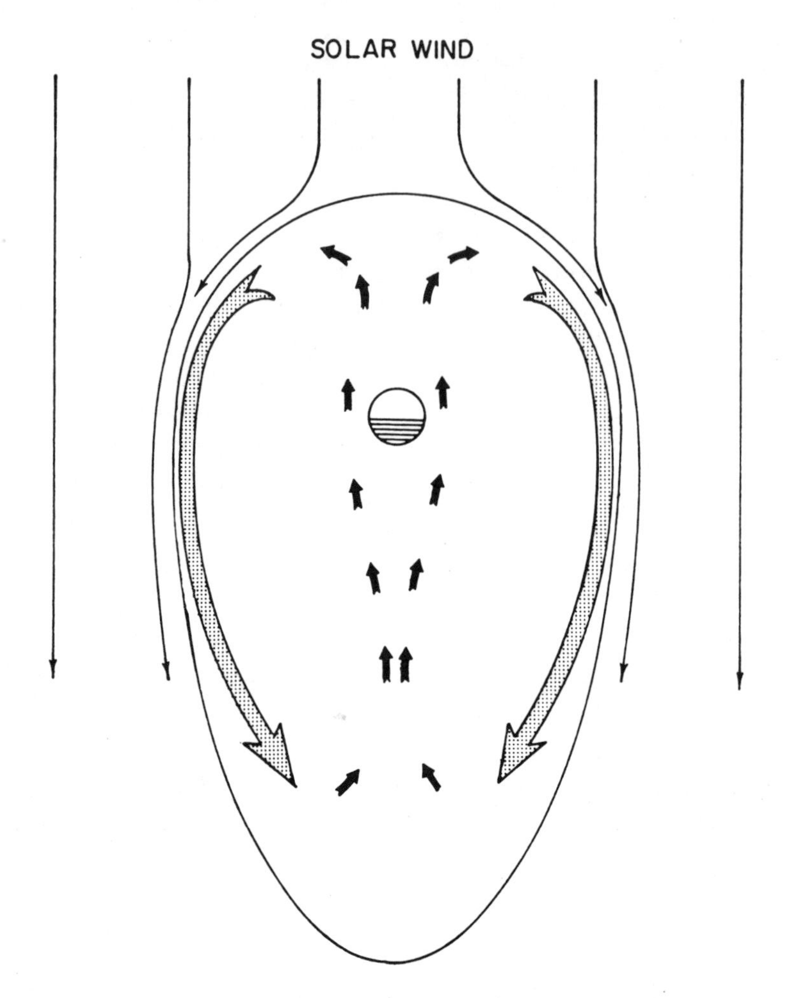

Figure 9.14 A sketch showing a cross-section of the magnetosphere in the equatorial plane depicting the convective motion of the outer magnetosphere that is imposed by the drag of the solar wind along the flanks of the magnetosphere. (After Ian Axford and Colin Hines.)

is to produce a slow drifting of auroral forms, generally toward the west during the evening hours, and to the east after midnight. The trend is not easy to see for someone watching the aurora, but it becomes obvious when a person looks at all-sky photographs of aurora played back at a speeded-up rate. The contraction of the sky image in the all-sky photographs combined with the contraction in time during the playback makes the persistent auroral motions easy to see.

Because charged particles flow freely along magnetic field lines, the geomagnetic field lines act like perfect conductors of electricity. Thus they form part of the circuit powered by the solar wind-magnetosphere generator, and they extend this circuit down to the ionosphere of the polar regions. In one fashion or another, the ionosphere acts as the primary electrical load on the solar wind-magnetosphere generator. The energy that this generator extracts from the solar wind causes the glow discharge we call the aurora; the energy also heats up

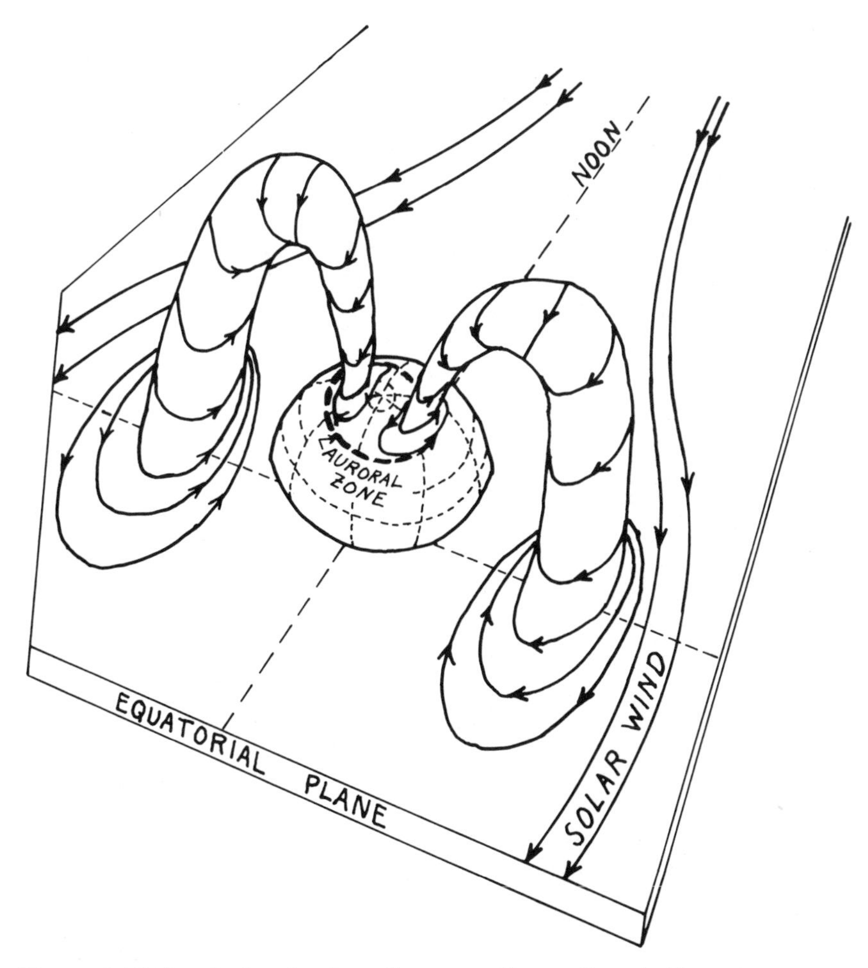

Figure 9.15 Sketch showing how the convection pattern in the outer magnetosphere maps down through the geomagnetic field to the auroral ionosphere.

and ionizes the high atmosphere and causes it to circulate.

Of the particle-borne energy in the solar wind that impinges on the front of the earth's magnetosphere, approximately 98 percent sweeps on around the sides of the magnetosphere and moves on out into the solar system beyond. The other 2 percent of the energy may be temporarily stored in the form of electric and magnetic energy, but it soon becomes kinetic energy carried by charged particles that either flow to the polar atmospheres to create the aurora or penetrate inward through the magnetosphere to enter the magnetic bottle known as the Van Allen belt, as illustrated earlier in Figure 9.12. As the particles stream through the magnetosphere they somehow get speeded up (accelerated) before arriving at the Van Allen belt or the polar atmospheres.

Uncertainty still remains about the details of how the merging occurs between geomagnetic field lines and those in the solar wind. Nevertheless, we now know that all or virtually all of the energy involved in the creation of auroras and related phenomena does cross into the magnetosphere from the solar wind.

Section 10
AURORAL CONJUGACY

In the very first textbook on the aurora, one published in 1733, the French natural philosopher Jean Jacques d'Ortous de Mairan suggested that auroras might occur in the southern hemisphere and that they might be similar to those observed in the northern hemisphere. Mairan thought that the aurora might be an extension of the sun's atmosphere that somehow mixed with the earth's. An external source of such a type, Mairan evidently reasoned, could lead to the aurora being similar in the two hemispheres. Mairan was thinking along the right lines, and he made a sound conjecture about auroral conjugacy.

European visitors to southern oceans, starting with Captain Cook in 1770, verified Mairan's conjecture that auroras might occur in the southern hemisphere. By the middle of the 1800s scientists were aware of a generally similar behavior of auroral displays in the two hemispheres, yet not until the late 1960s did they learn that the aurora in the southern hemisphere is essentially an exact copy (actually a mirror image) of that in the northern.

Section 10.1
THE MEANING AND IMPLICATIONS OF AURORAL CONJUGACY

The word conjugacy derives from the adjective conjugate which means paired, united or joined together. In the realm of geomagnetism, conjugacy refers to points on a given geomagnetic field line. Strictly speaking, any two points, regardless of their altitude above the earth's surface, are geomagnetically conjugate if they lie on the same geomagnetic field line. However, when people use the term 'geomagnetic conjugate points' they usually mean two points at the same altitude, and the altitude is often taken to be zero, so that the conjugate points are the two intersections of a geomagnetic field line with the earth's surface.

Merely a knowledge of the way charged particles like to travel along geomagnetic field lines poses an interesting question: if a person sees an auroral form over, for example, Fairbanks, Alaska, does an identical auroral form exist in the southern hemisphere? To answer the question we need to make an observation in the southern hemisphere at the appropriate location, the conjugate point to Fairbanks, because that is where the suspected auroral form should be. (This point is on the South Pacific Ocean, about 1500 km south of Christchurch and 600 km east of Macquarie Island.)

If we make a diligent search and the suspected auroral form is not found, then the implication is that the particle streams that generate the northern aurora are not intimately connected to those that create the southern aurora. But if auroral forms that appear over Fairbanks have identical or near-identical companions in the southern hemisphere, then the reason for the identity has to be that the causative particle streams are, at minimum, closely related. They may even be one and the same, a collection of charged particles that attaches to the looping geomagnetic field near the equatorial plane and penetrates down along the geomagnetic field to the atmospheres of both hemispheres.

In order to make observations of possibly conjugate auroras, it is necessary to identify the locations of conjugate points. The starting point is a numerical model of the geomagnetic field, a mathematical description of the field that can be used to calculate the direction and strength of the geomagnetic field at any location on or above the earth's surface. A person wishing to calculate the location of the conjugate to Fairbanks, Alaska, can instruct a computer to use the model to calculate the direction of the geomagnetic field at Fairbanks. The programmer then instructs the computer to go to a point a few kilometers in that direction, and again to calculate the direction of the field. By repeating this process thousands of times, the computer follows along the path of a geomagnetic field line out over the equator and back down to the conjugate point in the southern hemisphere, a journey exceeding 60,000 km. Just to make certain that the computation is correct, the programmer requires the computer to retrace its steps to see if it can get back to Fairbanks.

Section 10.2 OBSERVATIONS OF AURORAL CONJUGACY

Virtually all detailed information about auroral conjugacy comes from a series of 18 paired aircraft flights conducted by the University of Alaska and the Los Alamos Scientific Laboratory near the equinoxes in 1967, 1968, 1970 and 1971. The airplanes, long-range military versions of the Boeing 707, carried all-sky and narrow-field auroral cameras during the 1967 and 1968 flights and also sensitive auroral television cameras in the 1970 and 1971 flights. After taking off—one from Anchorage, Alaska, and the other from Christchurch, New Zealand—the paired aircraft flew carefully prescribed paths that allowed them to reach calculated geomagnetic conjugate points simultaneously, as depicted in Figure 10.1. Their routes took them poleward, across the auroral zones and back to the launching sites.

The pair of photographs shown in Plate 22, simultaneously acquired with narrow-field cameras mounted in the cockpits of the airplanes, illustrate the essential result: the auroral displays in the two hemispheres basically are identical mirror images, at least near the equatorward boundaries of the aurora. That is to say, a high degree of auroral conjugacy pertains, and the observations include instances of exact conjugacy.

Some differences do occur in the images acquired, and for several reasons. Although the aircraft were in communication and could adjust their speeds for winds, they did not always maintain conjugate positions. Even minor relative displacements of the airplanes sometimes caused significant differences in the appearance of the auroral forms observed. Also, the observations showed that the auroras in the northern hemisphere at this particular longitude, statistically (but not always) were brighter than their mates in the southern hemisphere. This difference can be explained on the basis of

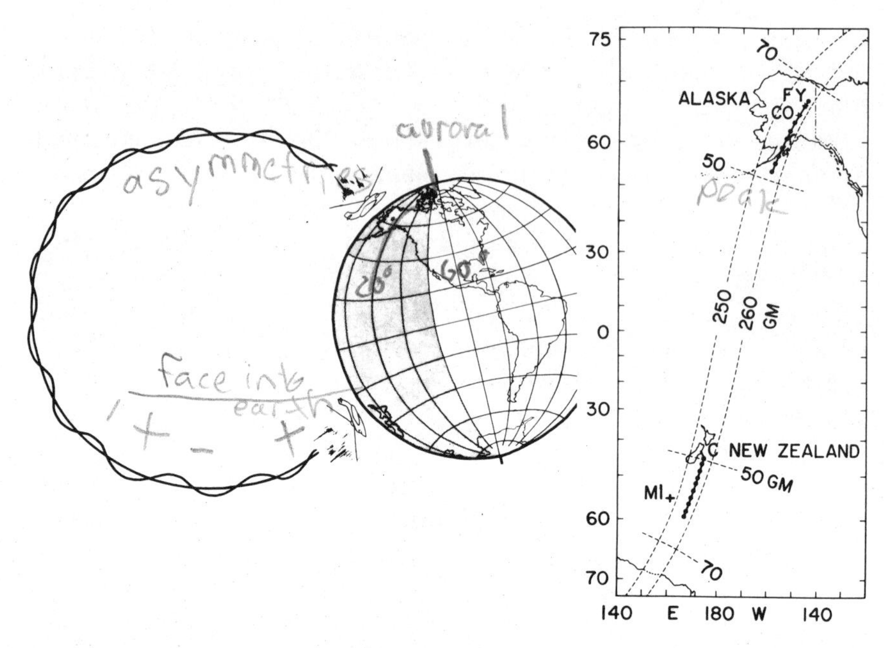

Figure 10.1 Illustration of the method of observing the conjugacy of aurora using aircraft flown from Alaska and New Zealand to arrive simultaneously at conjugate points. The map at right shows both geographic and geomagnetic latitudes and longitudes, on the map borders and along the flight paths, respectively. (Examples of conjugate points are the 10 pairs of dots on the heavy solid lines over Alaska and south of New Zealand.)

known asymmetries in the earth's geomagnetic field that focus more incoming particles into a given sky area over Alaska than over its conjugate area. Relative displacements from exact conjugacy also occurred. During periods of highest auroral activity, the southern auroras usually appeared closer to the equator than did their northern counterparts. Prior to auroral breakups, the auroras in the southern hemisphere usually lay somewhat westward of those in the northern, and after the breakup, eastward. These relative displacements increased toward the higher latitudes.

The general behavior is that expected if auroral activity creates some warping of the geomagnetic field. That the warping might occur is reasonable, because the incoming charged particle streams constitute electric currents, and the magnetic fields of these currents can distort the geomagnetic field to cause the observed displacements.

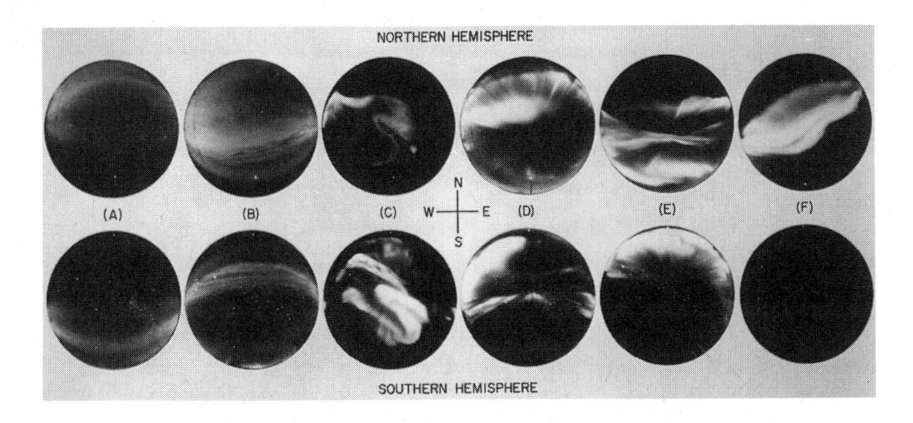

Figure 10.2 Pairs of simultaneous all-sky photographs acquired over the course of one flight mission across the auroral zones.

Sometimes at the highest latitudes and during the highest levels of activity, aurora was observed above one aircraft, but the sky was dark above the other. Either extreme displacements were occurring, or auroral conjugacy failed altogether. If it fails, then sometimes the interconnection between the two hemispheres provided by geomagnetic field lines is totally disrupted.

The sequence of all-sky camera pairs acquired on one flight (Figure 10.2) is typical. Pairs A, B and C portray a high degree of auroral conjugacy. The northern auroral forms in pairs D and E are displaced poleward of their southern counterparts, and pair F shows either a major displacement or a failure in conjugacy.

The photograph pairs shown in Figures 10.3 and 10.4 portray particularly good examples

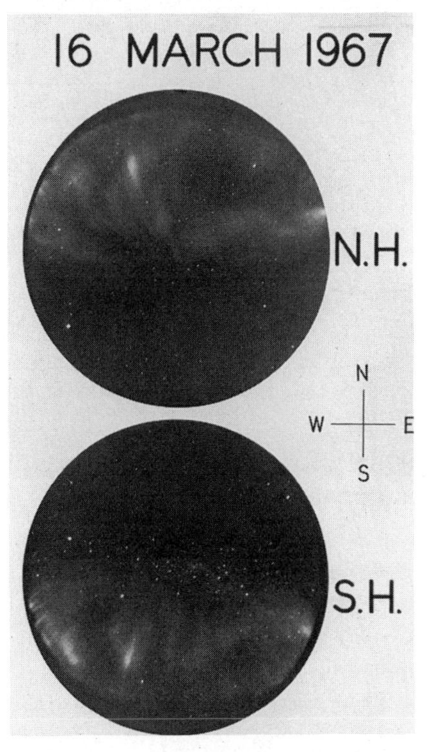

Figure 10.3 A pair of simultaneous all-sky photographs showing conjugate aurora. Note the bright ray-like feature that appears in both photographs.

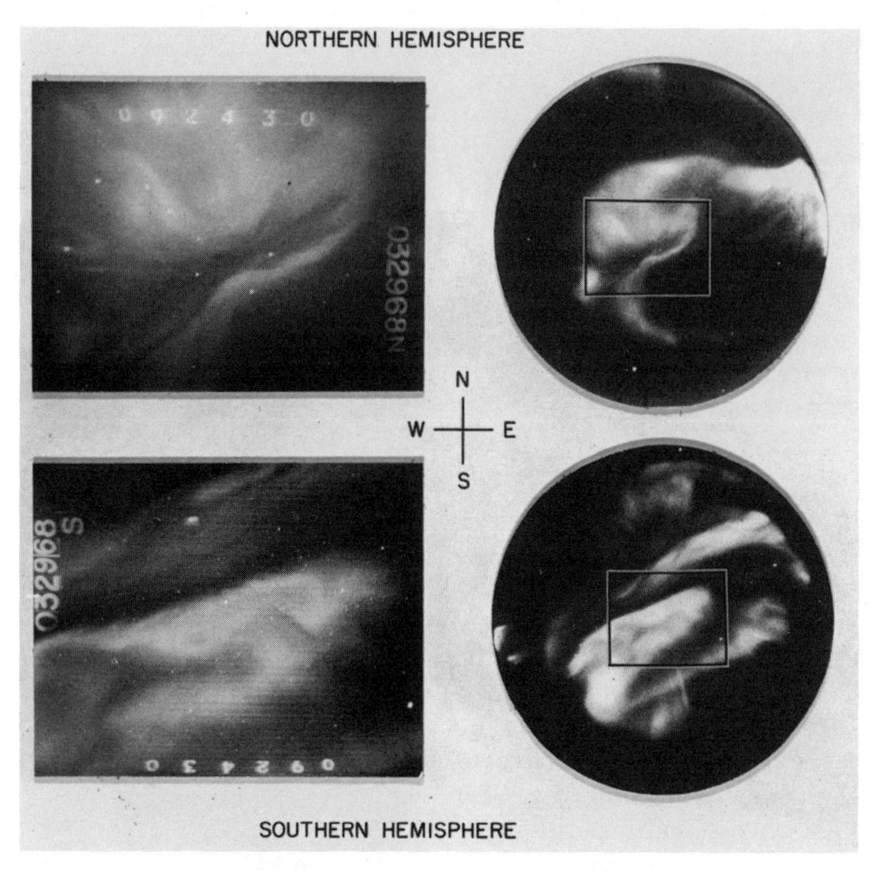

Figure 10.4 At left, television images of a conjugate looping structure. The boxes drawn on the all-sky camera photographs at right show the locations of the images at left.

of detailed conjugacy. Exactly conjugate rays appear in Figure 10.3, and Figure 10.4 contains a conjugate looping structure observed with both all-sky cameras and television systems.

The use of television systems on the aircraft led to one rather startling result. The systems detected pulsating auroras that were exactly synchronous; that is, the pulsating forms varied in brightness at exactly the same times in the two hemispheres within a tiny fraction of a second (See Figure 10.5). This result threw out one possible explanation of the pulsating aurora, that the pulses are caused by bundles of charged particles that move along the geomagnetic field lines so that they bounce back and forth between the two hemispheres.

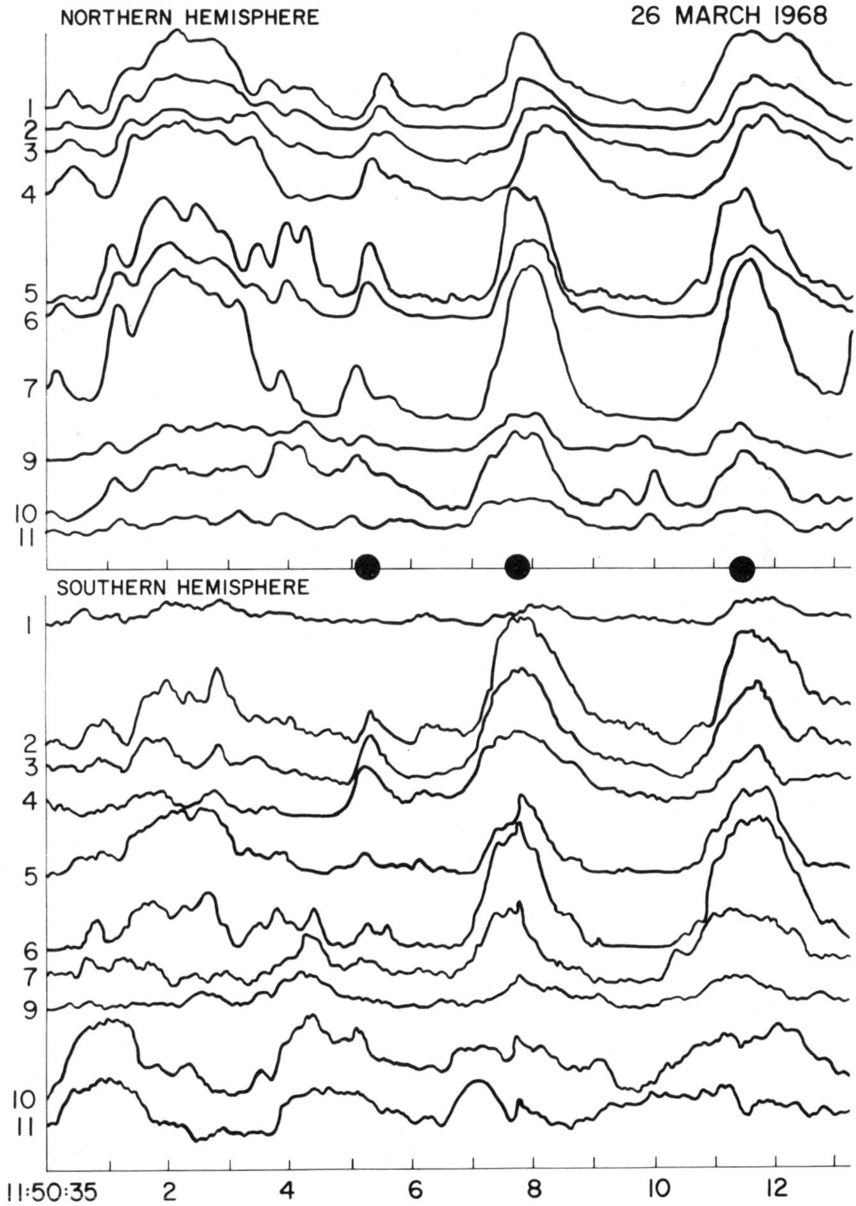

Figure 10.5 Photometer tracings of television recordings of pulsating aurora which show exactly synchronous pulsations in the two hemispheres. (Each tracing line represents a particular portion of the television image.) Note the prominent sychronous pulses at right, marked by black circles at center.

Repetitive bouncing of aurora-generating charged particles between the hemispheres is highly likely, since the charged particles contained in the Van Allen belt do undergo such motion. Those particles remain in the trapping region for long periods because their magnetic mirror points are above the atmosphere. Outside the Van Allen belt, in that part of the outer magnetosphere that carries the auroral particles, similar behavior is expected. Some of the particles headed into the high-latitude auroral atmosphere do not actually get there because they first reach their magnetic mirror points. They turn around and head back out across the equator and perhaps bounce several times before somehow getting lost through the walls of the imperfect magnetic bottle that pertains in this outer region. Those incoming particles that do not magnetically mirror can penetrate into the atmosphere to create the aurora. In the process, some of them rebound from impacts with the atmospheric atoms. By this process, called "atmospheric scattering," some of both the incoming primary particles and the secondary particles move upwards, back out along the geomagnetic field lines. These scattered charged particles then might cross the equator and penetrate into the opposite hemisphere a few seconds later to create aurora there. Thus it is quite reasonable to expect that pulsating aurora might be caused by the repeated bouncing of particle streams, and indeed some pulsations might have this cause.

However, the conjugate auroral flights revealed no examples of exactly out-of-phase pulsations (that is, alternating in the two hemispheres), so the observations strongly suggest that some other mechanism is required to produce the pulsating aurora. The fact that simultaneity can occur suggests an operative mechanism located in the equatorial plane, or at least a triggering there if the effective mechanism operates near or in the auroral zone atmospheres. That is one of the uncertainties; scientists still do not understand the relative importance of near-ionospheric and equatorial region magnetospheric processes in producing the aurora.

Section 11
THE CAUSES OF AURORAL MOTIONS AND SHAPE CHANGES

An inherent characteristic of the aurora is that it moves, and the individual auroral forms often change shape slowly or swiftly. It is possible to suggest mechanisms that will create all of the observed motions and shape changes, but additional mechanisms perhaps may be operating.

Section 11.1 AURORAL MOTIONS OF GLOBAL SCALE

Even if the aurora underwent no changes, an observer on the earth would see apparent motions due to the earth's motion beneath the auroral oval, and such motions are observed. The auroral oval behaves as though it it were a ring with off-center attachment to the geomagnetic pole, and it maintains a fixed orientation relative to the sun-earth line. Since the oval bulges out equatorward in the vicinity of the midnight meridian, the earth's rotation produces apparent north-south motions. For example, as Alaska swings beneath the oval during the evening hours, the oval appears to move southward. After midnight, it retreats northward. Also, if the auroral oval contained unmoving auroral forms, these nevertheless would appear to move westward, just as does the sun. The speed of this apparent motion is slow, only about 200 meters per second at the auroral zone (approximately the speed of a jet aircraft; at this latitude a jet can just keep up with earth rotation). Most actual auroral motions are much faster, so this apparent overall westward motion is barely noticeable and easily canceled by any eastward movements.

Another general motion of global scale is caused by the friction that the solar wind exerts on the flanks of the magnetosphere. The solar wind's attempt to pull the sides of the magnetosphere back in the antisolar direction generates a strong electromotive force that sets up two large convection cells in the outer magnetosphere. These extend inward to that part of the magnetosphere containing the charged particles that plunge into the ionosphere to create the aurora. (See Figure 9.14, previously presented.) The result is a slow drifting of auroral forms westward on the evening side of the earth, and a slow drifting eastward on the morning side, as depicted in Figure 11.1. The speed of the convection flow, as observed in

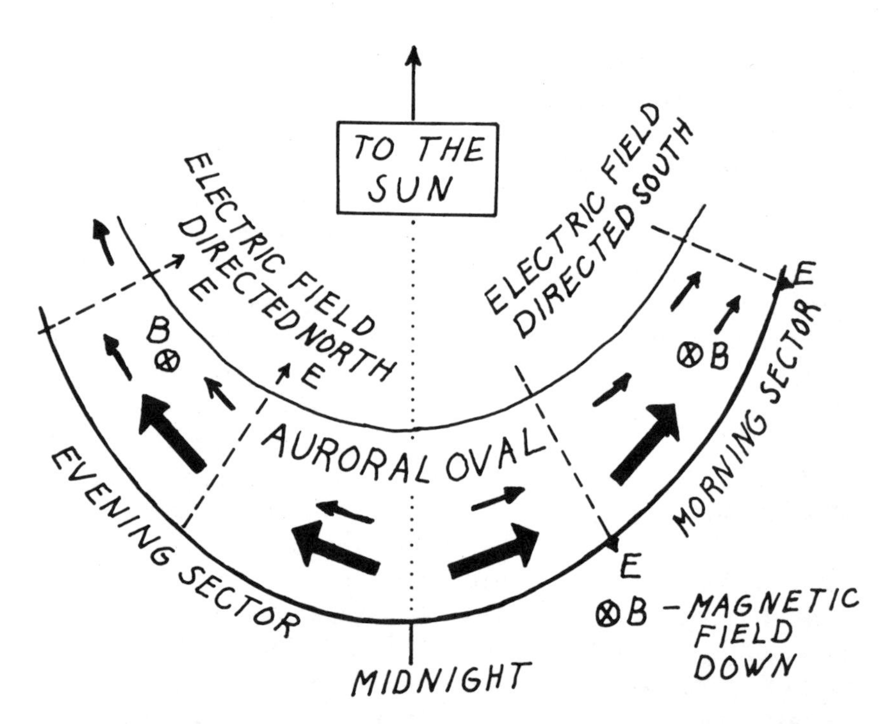

Figure 11.1 A sketch looking down on the oval from above, showing that auroral forms located in the evening sector tend to drift westward, and those in the morning sector drift eastward under the influence of magnetospheric convection. The convection carries down to the ionosphere through the action of the electric field, north-pointed in evening and southward in morning.

the aurora, is approximately 1 to 2 km/sec, fast enough to make this motion evident to the auroral observer who watches with care.

When the level of auroral activity increases, the diameter of the auroral oval increases and the oval widens. These changes cause an observer to see north-south motions that may be slow or sometimes very fast.

These broad patterns of motion are merely a matter of geometry, the viscous interaction between the solar wind and the outer magnetosphere, and the general level of auroral activity. They contribute little to the splendor of a fast-moving auroral display, although rapid changes in the width of the auroral oval can be spectacular.

Section 11.2 LARGE-SCALE AURORAL MOTIONS

One obvious cause of large-scale auroral motion is movement of the source regions in the outer magnetosphere, the regions where particles become attached to the geomagnetic field lines that guide them down into the auroral zone atmospheres. Presumably, the source regions are in or near the equatorial plane. Precisely how the particles enter the source regions or how pre-existing particles within the regions have their kinetic energies altered enough to become auroral primaries is unknown. But if such a region moves, the aurora produced by that region should also move. If the source region moves away from the earth, the aurora it produces should move northward (in the northern hemisphere), or if the source region moves inward toward the earth its aurora should move southward, as indicated in the top panel of Figure 11.2.

It is also possible to have large-scale motions of auroral forms even though their source regions in the distant magnetosphere remain fixed, because the geomagnetic field undergoes warping as illustrated in the lower panel of Figure 11.2. The incoming streams of charged particles constitute sheet-like electrical currents, and these currents have magnetic fields. The magnetic fields, although small, add to the pre-existing geomagnetic field at right angles and thereby slightly alter its direction on either side of the center of the moving charge sheet (see Figure 11.3). For this reason, if the incoming sheet that produces an auroral arc gets a slight irregularity in it,

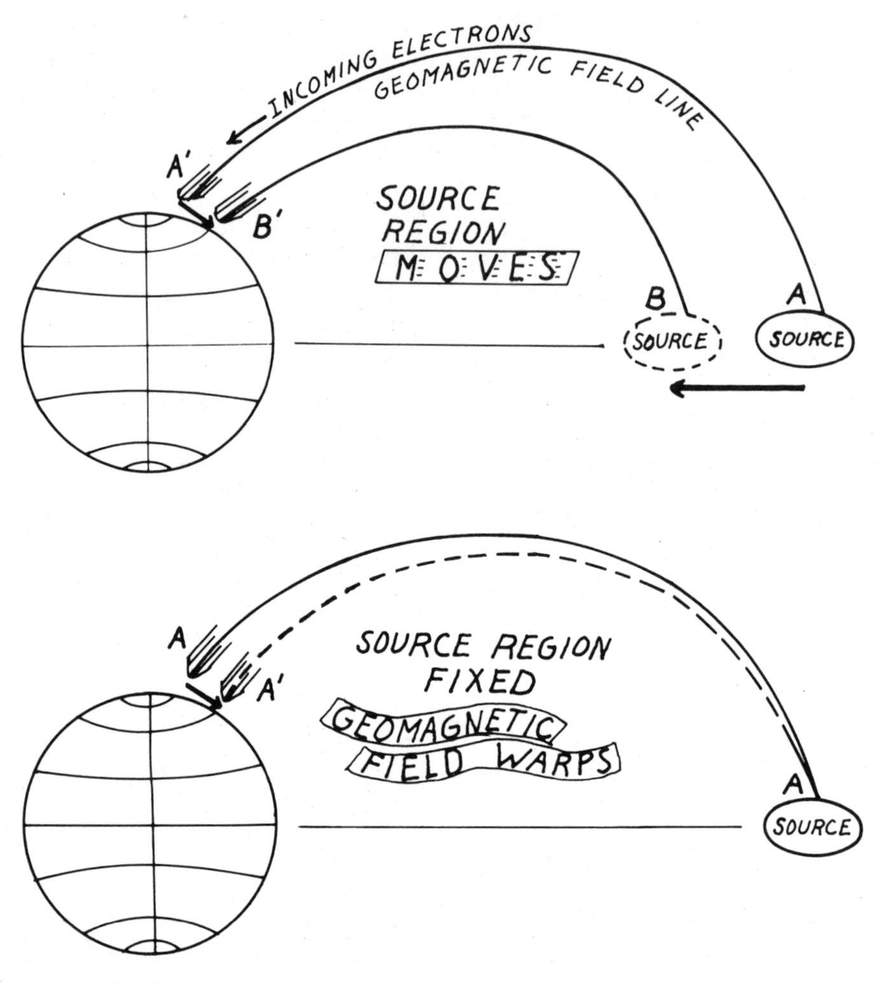

Figure 11.2 Sketches showing how (at top) an inward-moving source region in the magnetosphere's equatorial plane causes a southward-moving aurora in the ionosphere and, at bottom, how warping of the geomagnetic field can cause similar motion without movement of the source region.

the irregularity can grow. As it does, the sheet becomes distorted because the slight warping of the direction of the geomagnetic field carries incoming particles to new locations in the auroral atmosphere, as is indicated in the left-hand panel of Figure 11.4. The diagram shows how an auroral arc can become distorted by this means, first developing a fold and then perhaps progressing into a complex spiral shape. Spirals as

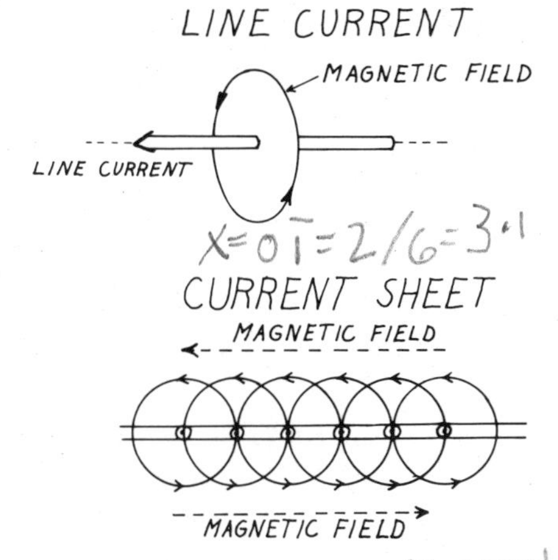

Figure 11.3 The magnetic field of a line current is circular about the direction of the current, as shown at top. A sheet current is, in essence, composed of an infinite array of line currents, the magnetic fields of which add to create a magnetic field oriented in opposing directions on the two sides of the sheet.

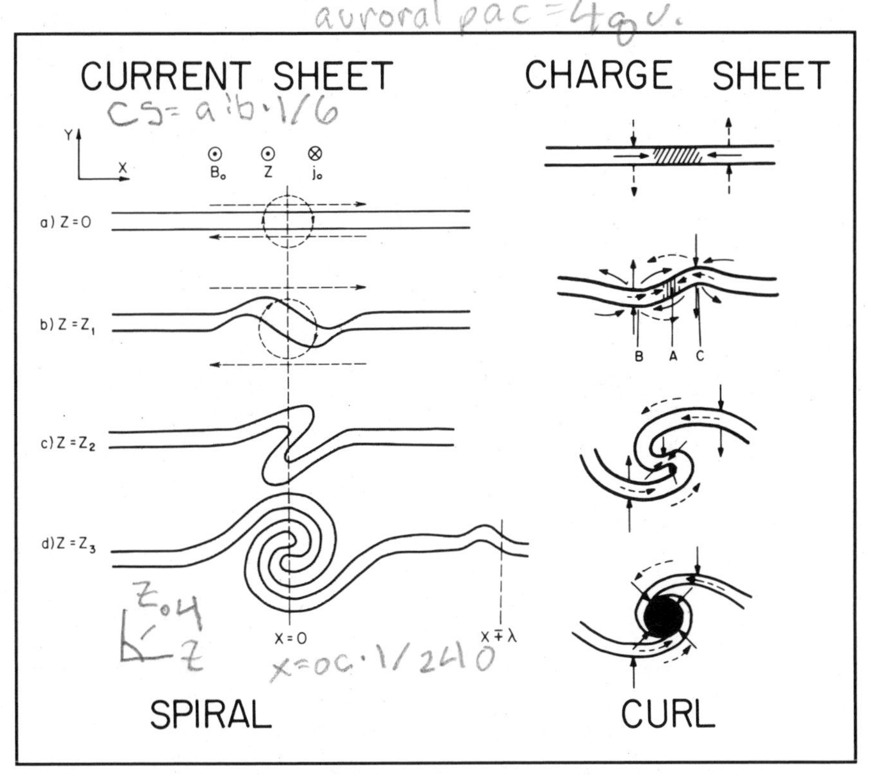

Figure 11.4 The perturbation of the geomagnetic field by the field of a sheet current tends to warp the sheet into a vortex, as shown at left. The result is the large spiral which grows from what was initially an auroral arc. A similar development caused by excess electric charge creates auroral curls (rays) as shown at right.

large as 1500 km across have been observed, implying that during their growth the geomagnetic field has been warped enough to displace portions of the original flat sheet as much as 750 km at its intersection with the atmosphere. Since the warping progresses downward from the equatorial plane, the current sheet there may have retained its original flat shape, as is suggested by the drawing in Figure 11.5. The mathematics of this warping process, a version of what is called the Kelvin-Helmholtz instability, is well enough known to permit calculation of the current required to produce the growth of a spiral. The calculated current is of an amount compatible with the observations of how many charged particles are entering the atmosphere to create the aurora.

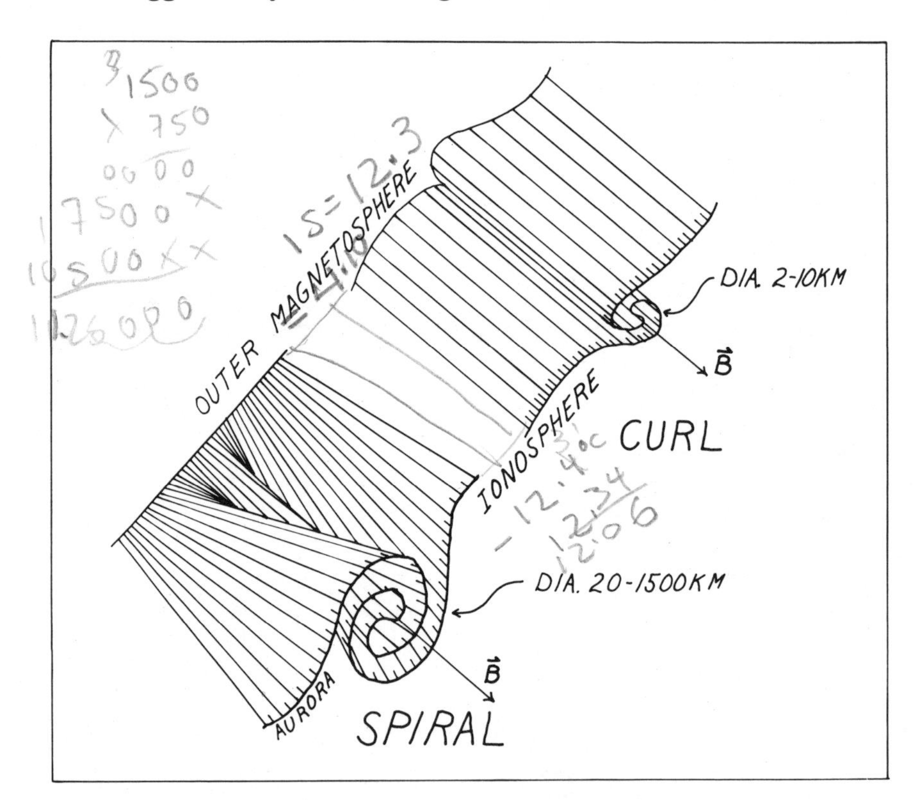

Figure 11.5 A sketch illustrating the difference in size between spirals and curls, and that the formation of spirals is associated by a twisting of the geomagnetic field lines. When a curl forms, the geomagnetic field lines remain mutually parallel.

Spirals have been observed both to wind up and to unwind. They wind up as the current increases (and the aurora in the spiral tends to brighten), and they unwind as the current decreases, typically with concurrent lessening of the auroral brightness. Sometimes the warping of the geomagnetic field does not progress enough to form a spiral, and the initial arc merely develops large folds.

Near the time of the auroral break-up, extremely bright auroral arcs sometimes appear to step across the sky. Instead of merely moving over, an arc located in one place suddenly jumps to an adjacent location, perhaps skipping across the sky in several steps. A possible explanation for this phenomenon is that the normally tight linkage between the magnetosphere and the ionosphere may be undergoing disruption. Under usual circumstances, the geomagnetic field lines provide a tight tie between the two regions. Thus any motion of auroral forms seen in the ionosphere is reflected by corresponding motion in the magnetosphere above, all the way out to the equatorial plane—except for those motions related to the warping that leads to large-scale folding and the formation of spirals. However, both theoretical and observational evidence suggest that, at altitudes a few thousand kilometers above the ionosphere, a special kind of turbulence can occur if the current in the incoming beam of auroral particles reaches a certain threshold value. If that happens, the tight tie between the ionosphere and the magnetosphere fails. It is likely that the stepping of auroral arcs across the sky, usually in the poleward direction, is a consequence of this temporary decoupling.

Section 11.3 SMALL-SCALE AURORAL MOTIONS

The rapid small-scale motions of auroras, such as the motion of rays along auroral forms, the formations of the rays (also called curls when observed in the magnetic zenith), and the development of small contortions of other types, appear to be due to the presence of electric fields that have components perpendicular to geomagnetic field lines. An electric field perpendicular to the geomagnetic field causes charged particles to move in a direction perpendicular both to itself and to the direction of the geomagnetic field. Near the northern auroral zone, the geomagnetic field points almost vertically downward, and there a southward-directed electric field causes both positively and negatively charged particles to move eastward. A northward electric field moves them westward. (For further details, see the discussion in Section 9.) Observations have shown that auroral arcs are accompanied by electric fields that point inward toward the center of the arcs. These electric fields are additive, so an arc that lies close to others comes under the influence of the combined electric fields of its neighbors. For this reason, an observer sometimes notices that the motions in an array of close-spaced arcs are as shown in Figure 11.6. The fastest motions are in the two outer arcs because those two are experiencing the largest combined electric field.

If for some reason the electron particle stream responsible for producing an auroral arc becomes slightly more dense in one region, the arc is likely to become contorted, and a ray (a curl) may develop there. The right-hand panel in Figure 11.4 shows the process, one very similar to spiral formation and leading to essentially the same final configuration—but one of far smaller size and of opposite sense of rotation. The region containing excess electrons is the center of an inward-directed electric field, as shown by the solid arrows in the right-hand panel of Figure 11.4. That field drives adjacent portions of the arc away from the axis of the arc, and if the process continues, the arc eventually winds up on itself to form a tight curl—an auroral ray seen end-on. 'Eventually' in this case is a short time, because the entire curl formation can occur in a fraction of a second, so rapidly that the human eye has difficulty following the process.

Once a curl vortex like this develops, it tends to cause the formation of another curl a short distance away along the arc. The overall result is the development of an array of evenly spaced rays along the arc. If another arc is nearby, the rays come under the influence of its electric field and therefore move along the arc. Contra-streaming rays along an array of bright auroral arcs are beautiful to watch, especially if they are red-tipped. (Incidentally, the same process—the formation of vortices and their rapid motions—takes place in the gas-filled tube that powers a microwave oven.)

The careful observer may notice that the red tips of rays sometimes appear to be moving faster than the green portions of the rays above. As noted in Section 7, this is a real phenom-

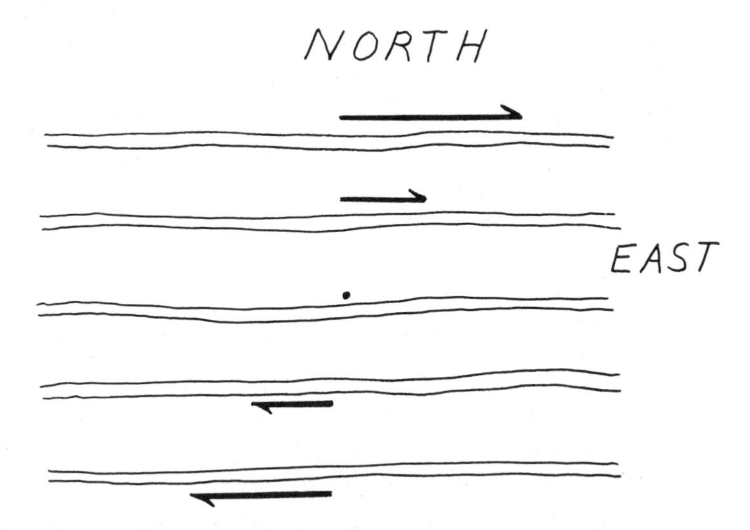

Figure 11.6 A sketch depicting an array of parallel auroral arcs as seen from above or below, showing opposing motions of irregularities on the two sides of the center of the array, and the higher speed of the motion toward the outside edges of the array.

enon created by the difference in the time taken by atmospheric atoms and molecules to emit the green and red light when struck by incoming electrons or protons. The red emission is instantaneous; it arises from oxygen and nitrogen molecules (it is not the oxygen atom 6300 red line previously discussed). However, the green emission (the 5577 A green line from atomic oxygen) is slower, on the average by 0.7 sec. Thus the red light shows the instantaneous location of the incoming particle stream, whereas the green light shows where the stream was located a fraction of a second earlier. Television observations of aurora made through color filters demonstrate that the green light in the aurora is somewhat smeared out compared to the red. Auroras observed in only red light are startlingly more distinct and more rapidly changing than those observed in green. Unfortunately for the auroral observer, human eyes respond best to aurora's green light.

The distortions to auroral arcs that may lead to the formation of curls (rays) sometimes do not proceed enough to cause more than a brief waviness in the shape of the arc. If the process continues a bit longer, then small, uniformly spaced folds occur, and sometimes the arc even splits into two separate sheets as the folds break apart. A Japanese scientist, T. Oguti, prepared the diagram in Figure 11.7 to illustrate some of the developments he observed. Part E of his diagram shows the formation of curls.

Section 11.4 SUMMARY

Auroral motions and changes in shape seem bewildering, yet they are orderly when examined in detail. Many of the changes in shape are caused by warping of the incoming particle streams by attendant magnetic and electric fields—a sort of self-induced boot-strapping operation, as it were. These changes, in one way or another, all involve what is called shear: differential motions within different parts of a fluid or gaseous material. Shear is common in nature. Shear eddies show on the surface of water in streams, or beside boats moving through water, in smoke curling up from a fire; they are responsible for clear-air turbulence and gusty winds. Shear is everywhere, even in the aurora, where one of its more spectacular consequences is the formation of spirals and curls (auroral rays).

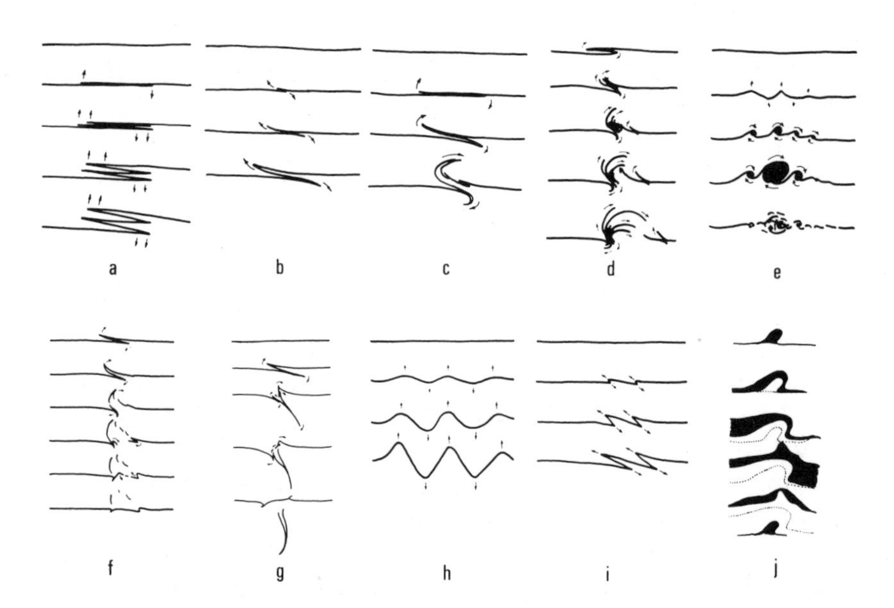

Figure 11.7 A diagram prepared by T. Oguti to illustrate various types of observed deformations of auroral arcs.

Section 12
ARTIFICIAL AURORAS AND CHEMICAL LOOK-ALIKES

Controlled experiments using electron beams to generate artificial auroras and certain chemicals injected by rockets and satellites into the high atmosphere or the magnetosphere have yielded valuable information about these regions and have verified hypotheses advanced on the basis of information gleaned from the aurora itself. A number of the chemical releases—some of which produce clouds similar to the aurora in appearance—have been at locations where millions of people could watch them, most notably on the Eastern Seaboard by means of rocket launches from Wallops Island, Virginia. The University of Alaska's Poker Flat Research Range near Fairbanks is another favorite location, and some releases from there can be seen almost everywhere in Alaska and northwestern Canada.

Section 12.1
ARTIFICIAL AURORAS

Proof that electron beams will travel great distances along geomagnetic field lines and will generate auroras when striking the atmosphere comes from several experiments conducted during recent decades.

The first controlled experiment successfully to generate man-made aurora took place at NASA's rocket launching facility at Wallops Island, Virginia, in 1969. (Before this experiment, the United States had produced artificial auroras in an uncontrolled fashion by exploding nuclear weapons in the atmosphere. Debris from the detonations had interacted with the atmospheric constituents to produce ray-like artificial auroras that were easily seen and photographed.) The Wallops Island experiment involved placing an electron accelerator in a rocket payload that was flown to an altitude of several hundred kilometers. When the payload reached altitude, it was tipped over so that the electron accelerator would fire short pulses of electrons downward along the direction of the local geomagnetic field. Each pulse lasted up to 1 sec and consisted of electrons with energy approximately 10 keV (kilo electron-volts), roughly the energy known to be carried by the electron beams that create the natural aurora. Several of the beams travelled from the rocket downward into the atmosphere a distance of 170 km and there produced ray-like auroras. These were too weak to be seen with the human eye but were detectable with highly sensitive television systems located in the general vicinity of Wallops Island. One TV image of the artificial aurora is shown in the left-hand portion of Figure 12.1.

The group performing the Wallops experiment later, in 1972, conducted a similar experiment in Hawaii that employed a rocket-borne electron accelerator intended to send electron beams upward. Equipment malfunctions interfered, but as the accelerator reached an altitude of 400 km over Hawaii, it ejected one pulse in the right direction. The pulse travelled upward along a geomagnetic field line, out over the equator, and back down into the atmosphere of the southern hemisphere at a location near that predicted. It travelled the 7000-km distance in the expected time, 0.2 sec, and generated the aurora seen at right in Figure 12.1. The aurora had a brightness and altitude that indicated no degradation of the electron beam as it transited the inner magnetosphere.

A group of scientists led by John Winckler of the University of Minnesota obtained even more impressive results with several rockets launched from Poker Flat, Alaska, in the 1970s. The rockets sent electron beams upward to travel along the geomagnetic field to the conjugate area where they mirrored and came back to the sky over Alaska about 2 sec later, travelling more than 200,000 km round trip. Instruments on the rocket payloads observed some of the beams, and a television system at Poker Flat detected artificial auroras generated by two of the beams. These experiments verified that electron beams lose little energy as they travel hemisphere to hemisphere along very

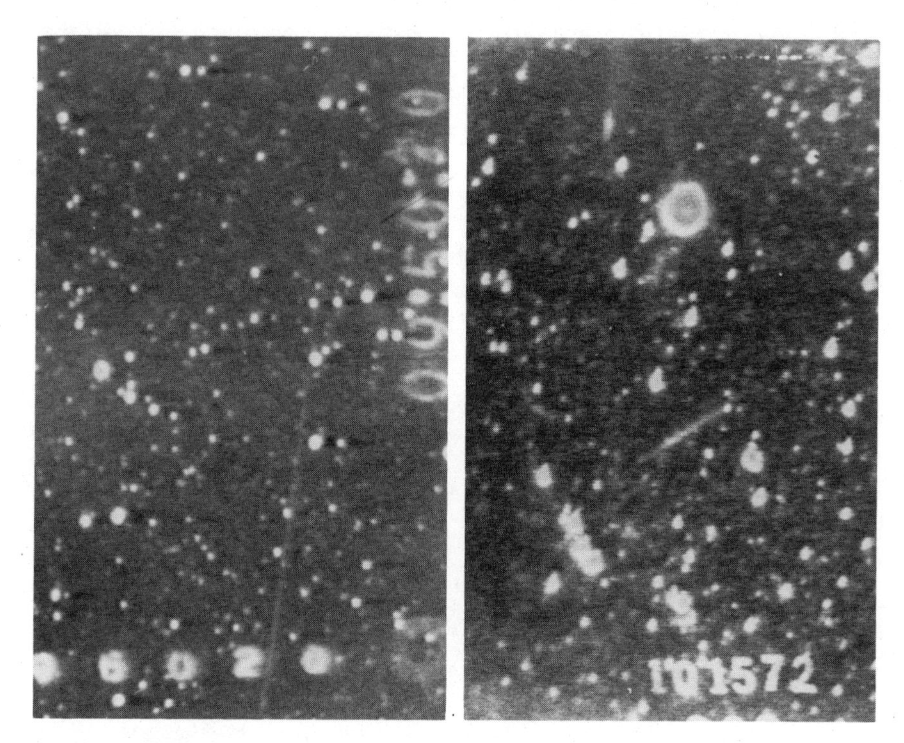

Figure 12.1 At left, a television image of the first controlled artificial aurora, generated by an electron beam fired down into the atmosphere over Wallops Island, Virginia, in 1969. The much shorter streak in the image at right is an aurora in the southern hemisphere near Samoa that was generated by an electron beam fired upward from a rocket flying over Hawaii in 1972.

long geomagnetic field lines, and that the shapes and lengths of these long field lines is that predicted by calculation.

Section 12.2
CHEMICAL TRACERS AND AURORA LOOK-ALIKES

Beginning in 1955, scientists began using chemicals injected into the high atmosphere as tracers that allow measurement of winds at high altitude. The first experiments involved releasing sodium vapor from rocket-borne canisters. Each canister carried approximately 1 kg of sodium metal in small pellets interspersed within finely powdered iron oxide and aluminum. When ignited with a hot wire, the iron oxide and aluminum chemically interacted to produce heat great enough to vaporize the sodium and eject it from the canister. The result was a visible yellow cloud created by the action of sunlight on the vaporized sodium atoms through resonance scattering, the process that contributes to the blue color of sunlit aurora. This cloud took up the motion of the surrounding air so that photographs of it allowed measurement of the wind speed and direction.

It is necessary to perform such wind tracer experiments during evening or morning twilight because at these times the sun can shine on the released sodium clouds, but the sky is dark enough that they can be seen and photographed. The method allows the measurement of winds in the altitude region 50 km to 250 km, a region where the winds sometimes are fast but the air is too tenuous to permit measuring the wind by other means such as employing falling parachutes or falling ra-

Figure 12.2 Photograph of a luminous sodium and lithium vapor trail over Wallops Island, Virginia, in 1969.

dar-reflective chaff. The wind typically varies in direction with altitude, as is readily seen in Figure 12.2 where the wind has contorted a trail of sodium and lithium vapor into a spiraling shape. In this photograph the portion of the trail first released shows globular protrusions, and the last part is at lower left where the trail produced by the now-falling rocket is still straight.

Sodium vapor placed in the high atmosphere becomes visible because it re-emits sunlight through the process called resonance scattering. In resonance scattering, an atom or molecule absorbs energy from impinging light of certain wavelengths and then re-emits the energy at the same wavelengths. The internal character of the atom or molecule determines the wavelengths of light involved.

Searching for brighter and more long-lasting tracer clouds, scientists have employed materials other than sodium. These include barium, cesium, lithium and strontium. In the process, residents living near rocket launching facilities have occasionally been startled by the sudden appearance of strange colors in the night sky, including the deep red produced by lithium.

Chemiluminescence, the process used by fireflies to generate light, also has allowed the use of certain compounds that can be released from rockets to create wind tracer clouds or trails visible at times other than twilight. For example, a mixture of cesium nitrate, aluminum and magnesium will react to produce a self-luminous trail lasting up to several minutes, and a release of trimethyl aluminate does even better.

The most spectacular tracer clouds are those produced by resonance scattering of sunlight from releases of barium vapor at high altitudes. These have high scientific value because they allow observations of winds and also of any electric fields that may be present in the high atmosphere. A finely divided mixture of barium metal and copper oxide, when ignited with a hot wire, will burn intensely (as is illustrated by the ground test pictured in Plate 23), and the reaction produces enough heat to vaporize and eject barium atoms. A canister containing only one kilogram or so of this mixture will produce a cloud 10 km across within a few seconds in the upper reaches of the atmosphere. Timed releases from multiple canisters carried aboard one rocket allow placement of the clouds at several altitudes, typically in the range 140 km to 400 km. Figure 12.3 illustrates the method (see also Plates 24 and 25).

A release of barium in the sunlit high atmosphere during twilight actually produces two clouds. One cloud is at first multi-hued (yellow and red) and then later basically green, while the other is initially purplish and later blue. The green cloud results from the excitation of neutral barium atoms by sunlight to create a strong resonance emission at 5535 A. This cloud grows into a spherical shape that drifts with the wind so that observation of its motion allows measurement of the wind speed and direction. More interesting is the blue cloud that develops (see Plate 24). It occurs because sunlight also ionizes some of the barium atoms which then emit strongly at 4934 and 4554 A. This blue barium

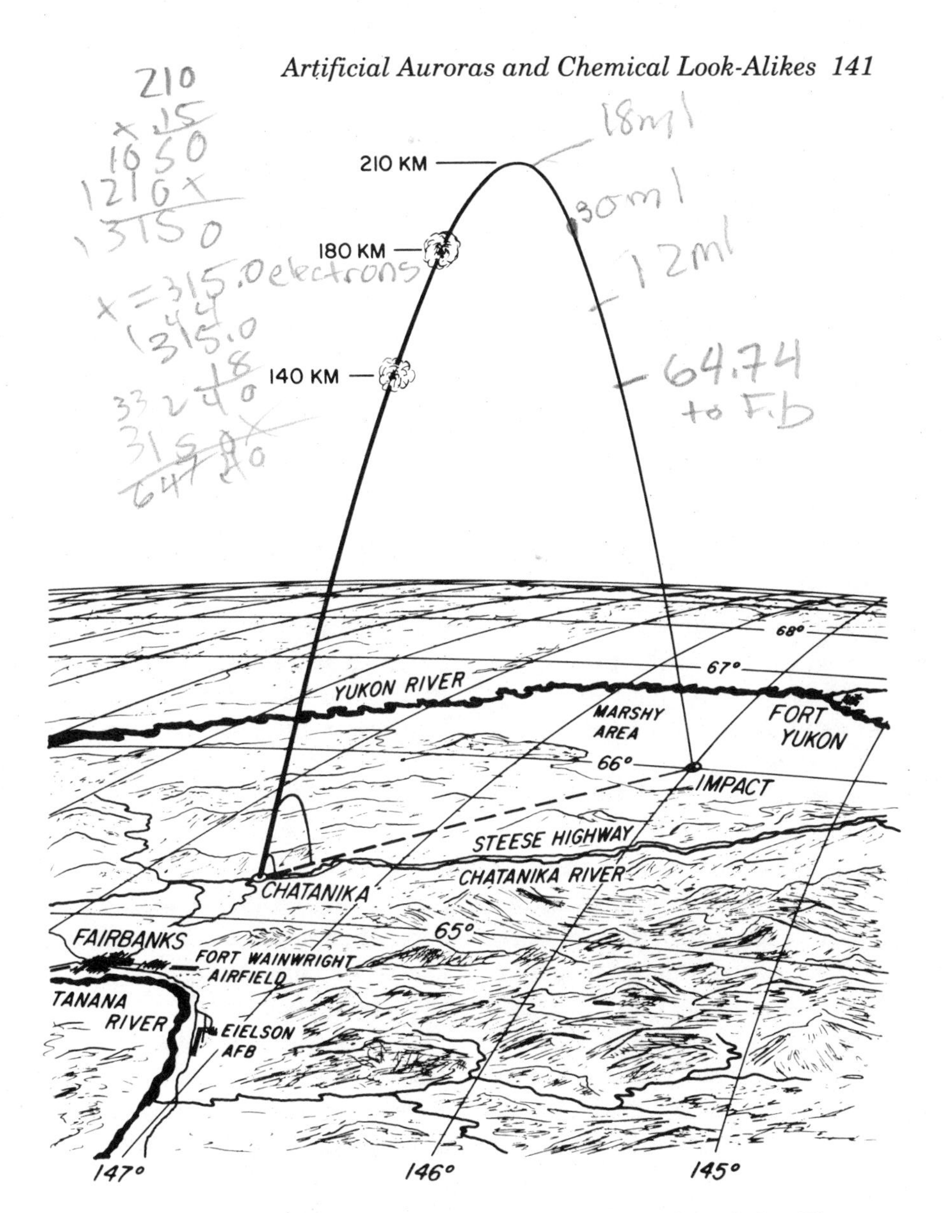

Figure 12.3 A drawing to illustrate the flight of a rocket from Poker Flat, Alaska, carrying chemical release canisters ignited when the rocket reaches altitude of 140 and 180 km.

ion cloud is readily distinguished from the green barium neutral atom cloud by its different color, shape and motion. Whereas the green neutral cloud grows spherically, the blue ion cloud quickly becomes cigar-shaped. Because they are electrically charged particles, the barium ions cannot move across the geomagnetic field, but they can move freely along its direction. So the blue cloud elongates in the direction of the geomagnetic field lines, essentially vertical at high latitudes. If an electric field is present, it acts in combination with the geomagnetic field to cause the blue ion cloud to drift horizontally at a speed and direction determined by the strength and direction of the electric field. Herein lies the power of this tracer technique: observation of the motion of the blue cloud provides an accurate means of measuring electric fields in the high atmosphere.

Typically, the barium ion clouds also stretch out horizontally in their direction of motion and become striated so that they look like rayed aurora, as Plate 24 indicates. If aurora is nearby or on the same geomagnetic field lines as the ion cloud, the motions seen in both are similar or identical, and so, except for its color, the barium ion cloud does indeed look like an aurora.

Most of the barium releases to date have been in the atmosphere, but others have been deployed in the magnetosphere high above to examine electric fields up there. Releases made even higher, near the outer boundary of the magnetosphere, have allowed attempts to follow the entry of charged particles through this boundary.

Although as little as 1 kg of barium-copper oxide mixture easily produces highly visible neutral and ionized tracer clouds, introduction of larger amounts of barium or other materials can change the local environment in the region of the release. Experimenters working in Alaska and elsewhere have sometimes deployed as much as 50 kg to see if such large releases would have significant effects. Little perturbation occurred, but the resulting clouds were spectacular: some spread over most of Alaska's skies, and even extended westward to the north of Siberia or eastward into western Canada. Certainly the perturbations were minor compared to that created by the Saturn rocket that lifted NASA's Skylab Workshop into orbit on May 14, 1973. As the Saturn passed through the ionosphere it released water and hydrogen molecules at a rate of over 1 ton per sec. These molecules reacted with the atmospheric gases to remove electrons so effectively that the launching created a hole in the ionosphere 2000 km across, and it lasted for several hours.

Figure 12.4 shows two views of an ion cloud produced by a

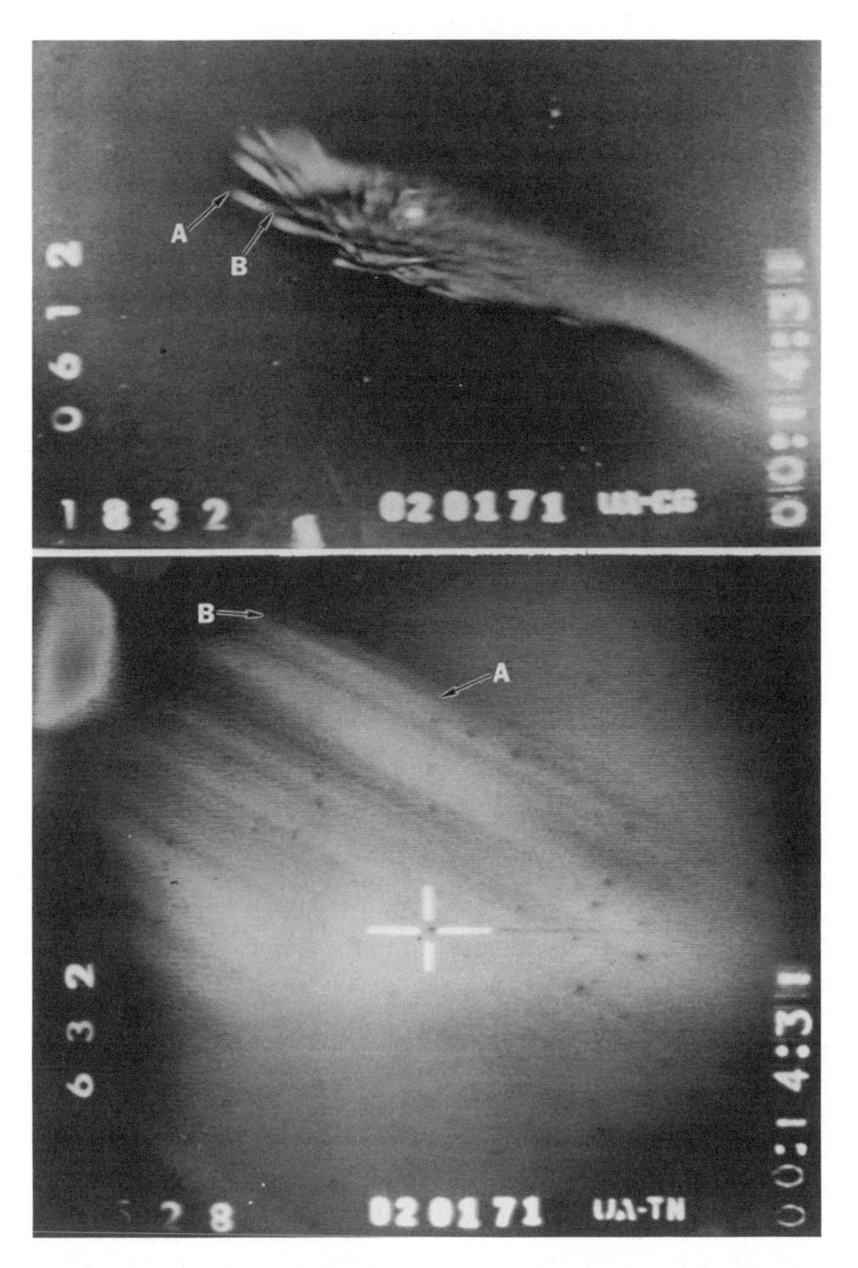

Figure 12.4 Simultaneous television images of a large barium ion cloud over Eglin Air Force Base, Florida, in 1971. The image at top is of the cloud near the magnetic zenith of one camera, and the one below is from a television camera located 100 km to the east. The letters A and B delineate a ray-like feature seen in both images.

very large barium release over Eglin Air Force Base, Florida, in 1971. The top panel shows the ion cloud in the magnetic zenith, while the simultaneous image in the lower panel is from a television camera located 100 km to the east. Notice the radical difference in the appearance of the cloud in the two images. The ray-like structure marked A - B is the same feature in the two images.

Section 12.3 MAGNETIC FIELD LINE TRACER EXPERIMENTS

A modification of the barium release technique permits the tracing of geomagnetic field lines over significant portions of their length, and even the full length of the shorter field lines that intersect the earth's surface at low latitudes. The technique uses shaped explosive charges to accelerate barium atoms to high speed, higher even than the speed needed for objects to escape the earth's gravity field, 11 km per sec. Antitank weapons (like the American and German bazookas used during World War II) operate in the same fashion. In these, a thin cone of copper is inserted into the hollowed-out end of an explosive charge. When detonated, the charge creates a shock wave that melts the copper and ejects it directionally out the axis of the cone at speed sufficient to penetrate several inches of armor steel and create a deadly inferno in a tank's interior. With better purpose in mind, scientists have replaced the copper with barium, as shown in Figure 12.5, and have performed the detonations high in the atmosphere where the air is thin enough not to impede the motion of the resulting barium jet. When the rocket carrying the shaped charge reaches altitude, its protective nose cover is ejected and the exposed cone oriented upward, generally along the direction of the geomagnetic field.

The detonation itself creates but minor ionization of barium atoms, but impinging sunlight ionizes many more. These ions then begin emitting light and fasten themselves onto the occupied geomagnetic field line and stream along it. Some of the ions travel as fast as 13 km per sec, and others move more slowly, so the overall jet smears out enough to make a long segment of a geomagnetic field line visible.

Figure 12.6 contains a series of television images of one of the first successful geomagnetic field line tracing experiments, one conducted by the University of Alaska

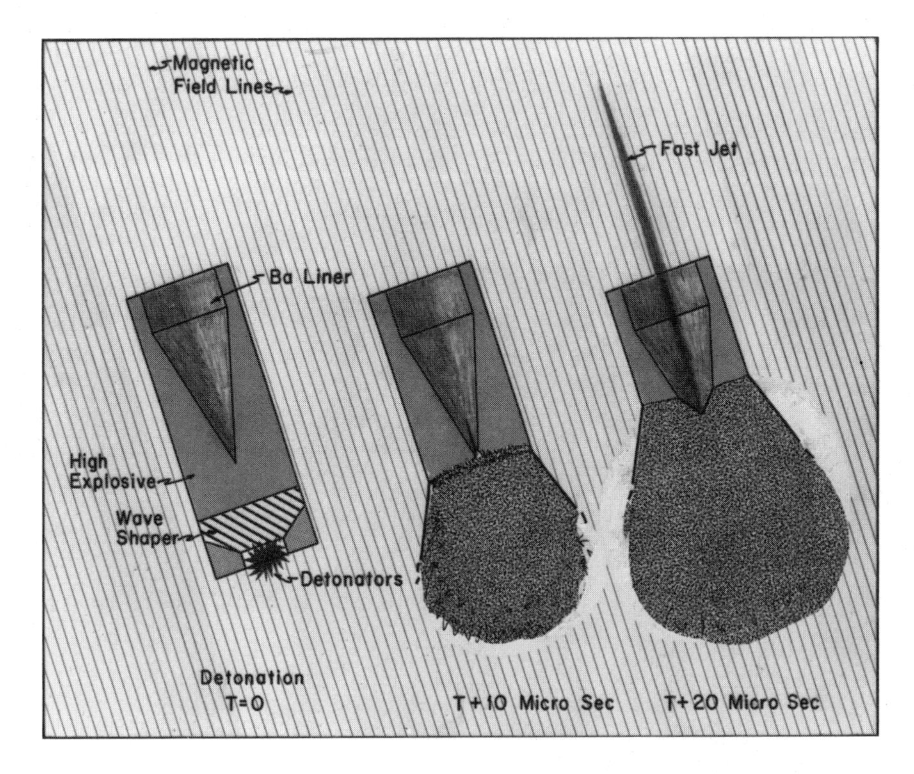

Figure 12.5 The barium shaped charge technique employs a high explosive to create a fast jet of barium ions that follow along the magnetic field and map its direction.

and the Los Alamos Scientific Laboratory over Hawaii in 1971. The sequence at bottom shows the appearance of the barium jet during the first seconds after detonation, as observed on a mountaintop several hundred kilometers distant. The top montage, from the same location and 4 min later, shows the barium jet streaming along the geomagnetic field line, making it visible along the northern 2000 km of its length. Central panels contain aircraft-based television photographs of the jet as it entered the atmosphere in the southern hemisphere 7 min after detonation, and later at 26 min. The experiment made the entire 7000-km length of the geomagnetic field line visible. Notice also that in the photographs taken 26 min after detonation, the single jet has split into several, making it look somewhat like a rayed aurora. While generally verifying the correctness of methods used to predict the locations of geomagnetic conjugate points, this

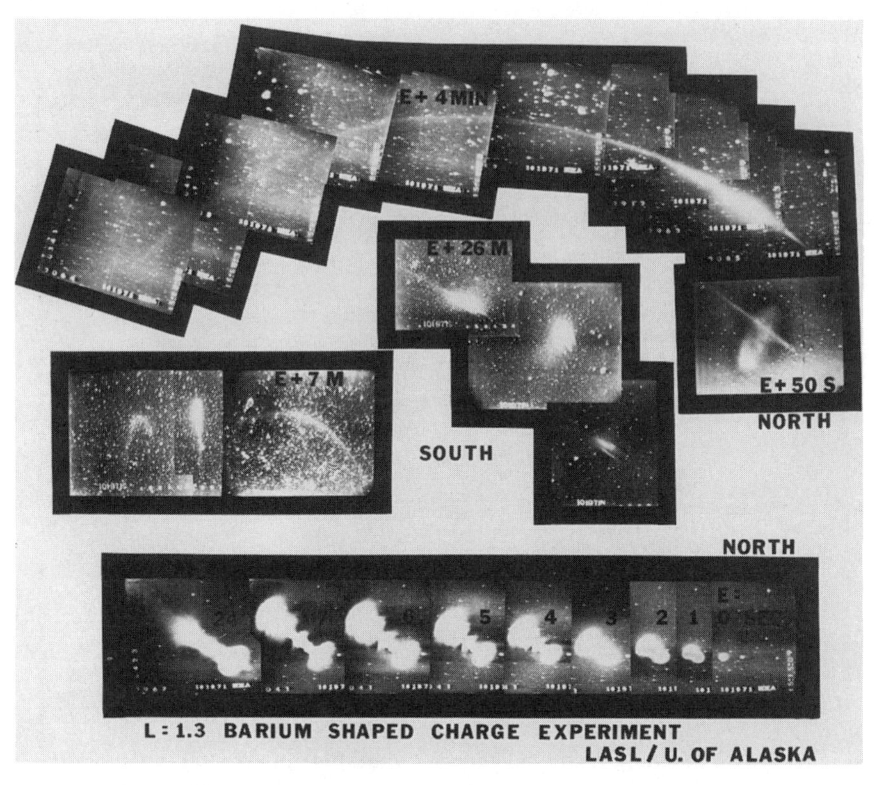

Figure 12.6 A montage of television images of a barium shaped charge experiment conducted over Hawaii (see text).

experiment led to some modification in the use of the models so as to obtain better predictions.

Figure 12.7 shows a barium jet streaking upward over northern Alaska as seen from an aircraft flying over central Canada. Note the wing of the airplane at lower left and, near it in the image, a band of aurora located far to the northwest of the aircraft.

Similar experiments conducted at the University of Alaska's Poker Flat Research Range have produced barium streaks that could be observed with ground-based television out to distances of 20,000 km, two-thirds of the distance to the equatorial crossing of the geomagnetic field line traced by the barium streak. In one case, a barium streak was detected near the equatorial crossing, 30,000 km away, by a photometer located in the state of Washington. Instrumented aircraft searched for the streak on its arrival in the southern auroral zone, some 3 hours after launch, without success.

Figure 12.7 A montage of television images showing a barium jet reaching up along the magnetic field 110 seconds after its ejection. The tip of the jet is at altitude 2,000 km; aurora shows below, as does the wing tip of the aircraft carrying the television camera.

The barium shaped-charge technique has proved itself a valuable tool for tracing geomagnetic field lines and also for measuring electric fields in both the ionosphere and the magnetosphere. Barium jets have provided solid evidence of electric fields aligned parallel to the geomagnetic field, an important finding because such a field accelerates the charged particles that cause the aurora. Scientists have long known that the acceleration must occur somewhere in the magnetosphere, and thanks partly to the barium experiments and to satellite observations, one accelerating region is now known to be located several thousand kilometers above the auroral atmosphere. Acceleration may also be occurring much farther out, at distances near 30,000 km where the geomagnetic field lines that carry auroral primaries cross the equatorial plane.

Section 13
AURORALLY RELATED PHENOMENA

The aurora is but one manifestation of a complex sequence of processes that begin in the sun, pervade the interplanetary medium and extend through the magnetosphere down into the earth's atmosphere. A complex sequence, yes, but each process in the sequence probably will appear comparatively simple by the time all are recognized and understood. Scientists have not yet arrived at one major goal they seek—a full understanding of the solar-terrestrial relationship. Nevertheless, recent decades have brought major advances and the hope that all pieces of the puzzle will eventually mesh into a satisfying picture.

When a person ponders over the pieces of information now available and where they fit in the overall scheme, it helps to think about an important aspect of the interplay between the charged particles and fields that determines which controls the other. The idea here is that the interplay is at least slightly akin to what happens when a wind sweeps across a forested countryside. When the wind is strong it dominates the trees—it tears away their limbs, bends them over and even perhaps topples them—all without much impact on the wind itself. Yet if the wind is weak, the trees dominate it—they impede its flow sufficiently that a person on the forest floor might be unaware that the wind is even blowing. Electromagnetic fields and charged particles have a similar interplay: where the fields are strong, they dominate the motions of the particles, but where the fields are weak, the charged particles dominate them.

Within the solar wind, the magnetic field is very weak, so weak that it can little affect the charged particles blowing out away from the sun in the stream. Indeed, within the solar wind, the charged particles essentially carry the weak magnetic field along with them. As some of these particles encounter and cross the boundary of the earth's magnetosphere, they enter a region where the magnetic field is stronger. The magnetic field here is powerful enough to impede the particles and affect their paths, yet not strong enough for complete domination. The charged particles try to sweep the geomagnetic field away and, in the process, create the magnetosphere's long tail extending away from the sun. Their motion also establishes a powerful generator of electromotive force that sets up a convective motion of the magnetic field lines in the outer magnetosphere and provides the energy to power the aurora and its related phenomena. Those particles that do penetrate through the outer magnetosphere encounter an increasingly stronger magnetic field, one now strong enough to provide major guidance. The geomagnetic field here steers the charged particles and diverts a major fraction of them into the polar regions where they descend into the atmosphere and generate the aurora. Another large fraction, however, penetrates more deeply into the magnetosphere, encountering a strong enough geomagnetic field to dominate the particle motion completely. The geomagnetic field's configuration here is that of a magnetic bottle, and so it traps the charged particles into orbits from which they cannot escape if simply left to themselves. Some do escape, but only because they occasionally run into something that affects their motion or causes them to lose the charge that gives the geomagnetic field its control over their motion.

Section 13.1 REPLENISHMENT OF THE VAN ALLEN BELT

As the above discussion suggests, one of the important aurorally related phenomena is the replenishment of the Van Allen belt during times when auroral activity is high. The Van Allen belt is that portion of the inner magnetosphere that constitutes a nearly perfect magnetic bottle. It is filled with a semi-permanent collection of charged particles, primarily electrons and protons somewhat more energetic than those that penetrate the atmosphere to generate auroras. These particles undergo a complicated three-part motion: Each particle gyrates about the direction of the geomagnetic field (several hundred thousand to 10 million times each second, depending on its energy). While gyrating, the particle moves along the direction of the geomagnetic field to bounce between mirror points in the northern and southern hemispheres (taking about 1/4 to several seconds for each bounce cycle, again depending on the particle's energy.) Third, the particle drifts longitudinally, toward the west if positively charged, and to the east if negatively charged. The time taken to drift around the earth, depending on the particle's energy, ranges from a few seconds to about 10 hours. This overall three-part motion combines to cause the particles to move around the earth on surfaces called drift shells that resemble the outer surface of a doughnut and that have thicknesses determined by the gyrational motion of the particles about magnetic field lines.

The drift motion of the charged particles, positives westward and negatives eastward, constitutes electrical current in the westerly direction. This current has a magnetic field equivalent to that of a ring current (a current loop) encircling the earth; it weakens the earth's geomagnetic field observed on the surface at low and middle latitudes. Interestingly enough, the resulting depression of the geomagnetic field observed on the earth is directly proportional to the kinetic energy contained by the charged particles in the Van Allen belt. Therefore, measurement of the depression with magnetometers yields a measure of the kinetic energy in the trapping region, and that is related rather directly to the population of trapped particles.

The relationship between particle populations in the Van Allen region with depression of the geomagnetic field dictates that the periods long known as magnetic storms—periods of one to several days during which the field is depressed—are periods when the population of charged particles becomes enhanced. Observations show that the enhancement occurs during substorms, the almost-explosive events that bring increased flows of particles into the auroral zones to create brilliant auroral displays. Thus, the enhancement of the Van Allen belt region is intimately related to occurrence of auroral displays. The two are children of a common parent, the complex and as yet not fully understood process or collection of processes referred to as the auroral substorm (and also the magnetic substorm).

The Van Allen belt acts very much like a leaky water tank. If no new water is coming into the tank, the tank runs dry, but if buckets of water are poured into it at a fast enough rate, the bucket fills. (This general idea leads to my diagram of the "Expandable Tippy Thundermug" model of the magnetosphere shown in Figure 13.1, which one wag (Ian Axford, editor of the prestigious *Journal of Geophysical Research* at the time) said was the only theory of the magnetosphere that held water.) Decay in the population of charged particles in the Van Allen belt proceeds slowly enough that several days must pass before most of the contents are gone, although as the population decreases, the rate of decay slows down, and the region never entirely empties. During magnetic storms, periods when substorms are large and occurring one after the other, the Van Allen belt fills toward maximum capacity. That filling may have

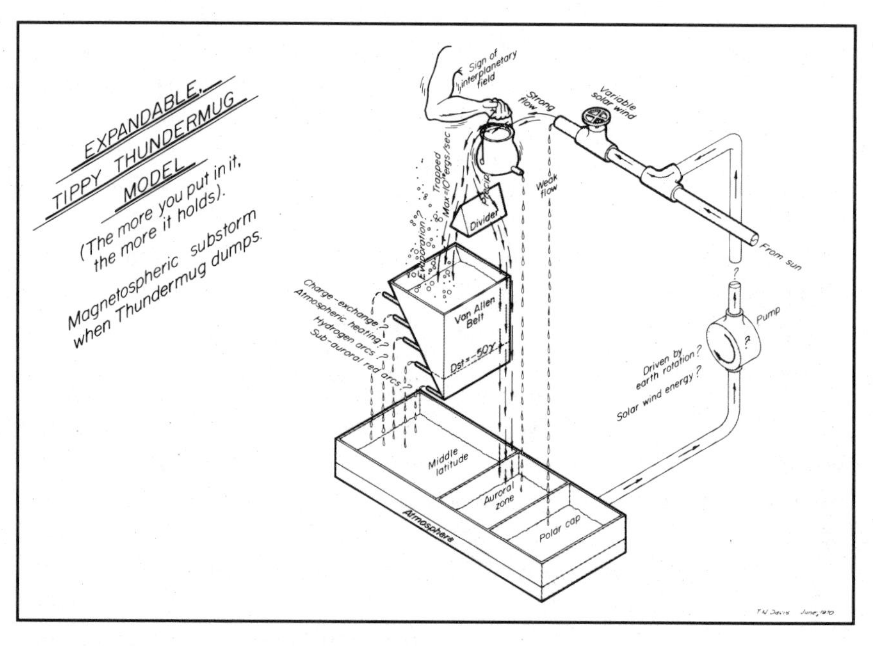

Figure 13.1 A semi-accurate portrayal of how the magnetosphere works in that it does not strongly enough emphasize the importance of the direction of the solar magnetic field carried in the solar wind.

some feedback on the aurora as well, because the increasing magnetic field associated with the ring current may be affecting the diameter and width of the auroral oval. That, however, is merely conjecture, and whatever the exact details of the relationship, it is clear that the association between the aurora and the Van Allen belt is intimate.

Not all of the charged particles in the Van Allen belt get there by substorm injection from the outer magnetosphere. Cosmic rays that enter the earth system strike oxygen and nitrogen molecules in the atmosphere and produce neutrons. Some of these diffuse upward into the Van Allen belt and there decay into electrons and protons. The Van Allen belt also loses its charged particles and kinetic energy to the atmosphere by at least two processes. One involves collisions of the charged particles with constituents of the outer fringe of the atmosphere, causing the charged particles to alter their motion enough to escape the trapping region. A highly energetic proton in the Van Allen belt also can collide with hydrogen, perhaps losing some energy, but mainly stealing away the electron from the hydrogen atom. That interaction converts the hydrogen atom to a slow proton, and makes the original energetic proton into a neutral hydrogen atom. It still carries abundant energy, but since it is now neutral it escapes the influence of the geomagnetic field, and hence the Van Allen belt region. The same number of protons remain, so the process causes the Van Allen belt to lose energy but not particle population.

Energy is the important thing, and much of it is involved in magnetospheric processes. The substorm injection events that bring energy into the trapping region may, in a few hour's time, deposit energy equal to that released by a Richter magnitude 7 earthquake or a 1-megaton nuclear weapon. Yet all that energy is carried by very little mass. If all the trapped particles in the Van Allen belt were assembled into a barbell, a strong man could pick it up.

Section 13.2
MODIFICATION OF THE HIGH ATMOSPHERE AND THE IONOSPHERE

Only approximately 4 percent of the energy carried into the atmosphere by auroral primaries produces auroral light. Most of the energy dissipates through heating the high atmosphere and through particle impacts that transfer kinetic energy to the resident gases and that break up the gases into component parts, some of which may be charged, i.e., ionized. The energy carried in by the auroral primaries makes profound local changes to the high atmosphere and the ionosphere; however, globally, sunlight is the more important agent.

Up in the high atmosphere, in the realm of the aurora, the atmosphere is quite different from that near the earth's surface. Near the ground, atmospheric gases are molecular, and tightly packed. A molecule there is likely to run into another if it travels only about a millionth of a centimeter: in the language of physics, its mean free path is one millionth of a centimeter. Yet at altitude 100 km, near the base of the aurora, the mean free path is 10 to 100 cm. At altitude 1000 km, the mean free path is near 200 km. The mean free path is directly related to the density of particles. Near the ground the density is approximately $10^{19}/cm^3$, but at altitude 300 km the density is 10 billion times less, only about $10^9/cm^3$. That low density, however, far exceeds the density of particles in the solar wind: roughly $5/cm^3$.

Another altitudinal difference is in atmospheric composition. Sunlight impinging on the atmosphere within a few kilometers of the earth's surface provides enough energy to the lower atmosphere to cause turbulent motions that keep the atmosphere mixed and of uniform composition up to approximately 100 km. Above that altitude, the constituents sort themselves out under the influence of gravity, the lightest ones rising to the top. Sunlight affects the composition there as well by breaking up molecules into atoms so that at higher altitudes atoms predominate over molecules.

Only slightly less than half of the sunlight incident on the high atmosphere actually reaches the earth's surface. Approximately 34 percent reflects or scatters back into space, and 19 percent is absorbed by the atmosphere. Most of the absorption (about five-sixths) is by water vapor and carbon dioxide in the lower atmosphere. The remaining one-sixth is ultraviolet light that is absorbed either high in the atmosphere or in the ozone layer at altitude 30 km. In essence, the ozone layer grabs up nearly all of the ultraviolet sunlight that was not absorbed higher up, allowing little to reach the earth's surface. Ultraviolet photons are highly energetic; in addition to burning sunbathers on the beaches, they are capable of destroying molecular residents of the high atmosphere.

Sunlight's ultraviolet photons carry enough energy to dissociate the nitrogen, oxygen and other molecules of the high atmosphere into their constituent atoms. The photons also are energetic enough to ionize the various molecules and atoms, thereby creating a population of positively charged atoms and molecules and negatively charged electrons. The consequence of this ionization is the ionosphere that encircles the entire earth in the altitude range 50 km to above 400 km. The ionosphere actually is a collection of several layers, each the product of ionization of one or more resident species by the incoming solar ultraviolet light. Actually, within the ionosphere the population of ionized particles is but a small fraction, far less than 1 percent, of the total particle population. Almost all of the particles in the region we call the ionosphere are electrically neutral, but the minute ionized fraction has profound effect on the behavior of the region.

The lowest of the ionospheric layers is the D Region with peak ionization density at about 60 km. Ionization of nitric oxide is its primary cause. Once formed, the D Region ionization quickly decays because the overall density of the atmosphere at this altitude is great enough to create frequent collisions of the electrons and positive ions with other constituents in reactions that recombine the electrons and positive ions back into neutral molecules. A consequence is a radical day-night difference in the concentration of electrons and positive ions within the D Region. When the sun sets on this layer, the concentration drops by more than a factor of 10.

Above the D Region lies the E Region, the product of ionization of oxygen and nitrogen molecules at altitudes near 110 km. The concentration of electrons and positive ions in the E Region is about 100 times greater than in the D Region and, as in that lower layer, the concentration changes ten-fold from day to night.

Higher yet is the F Region, a thick layer with maximum ionization density near altitude 250 to 300 km. Ionization of oxygen atoms and nitrogen molecules produces most of the F Region ionization, most of the ionized particles drifting upward from below. The concentration of ionization in the F Region is generally greater than in the layers below, in daytime typically 10 times that of the E Region. At night, the overall F region ionization decays to about one-fifth its daytime value.

Each of the three ionospheric layers has a characteristic of special interest. The F Region is important to long-distance communication because it reflects radio waves sent up from the earth below. The D Region acts to hamper communication because when it is dense, it tends to absorb the radio waves that would have reflected from the F Region had they reached it. The E Region supports electrical current in the upper atmosphere, current that is diffuse but intense enough to produce variations in the geomagnetic field on the earth's surface.

Incident sunlight is the essential cause of the global ionosphere, and the ionosphere varies daily and seasonally in response to the amount of sunlight striking each part. Profound changes also occur as the result of ionization by incoming auroral primaries. As these particles impinge on the high atmosphere, they create additional ionization that adds to that produced by sunlight. The auroral zone ionosphere is chaotic by comparison to the ionosphere at lower latitudes. The F Region becomes more reflective to radio waves, and the D Region more absorptive. Radio wave communication in the frequency range (1 to 20 megahertz) normally employed for long-distance transmissions at low and middle latitudes becomes essentially nonexistent in the auroral regions during extensive auroral displays. Even the high-frequency signals used to communicate between satellites and the earth's surface are adversely affected, although not so severely. Processes associated with the precipitation of auroral particles also generate radio waves in a broad range of frequencies, and these can cause additional interference.

Auroral particle precipitation during substorms also enhances the density of ionization in the E Region and gives that layer a greater ability to carry electrical current. Equally important are the electric fields imposed on the ionosphere by convection of the outer magnetosphere during times of high auroral disturbance. Within the F Region the imposed electric fields act in conjunction with the geomagnetic field to move ions horizontally. The motion of negatively charged electrons and positively charged ions is in the same direction, so no net current results (the positive current cancels the negative.) In the

E Region, however, the density of the atmosphere is great enough to cause the heavy positive ions to drag much more against the neutral atmosphere than do the lighter electrons. Thus, the electrons move more rapidly, and the result is a net current. Although diffuse relative to the current carried by a wire, the current in the E Region close to auroral forms is intense. The current within and near the auroral oval is so much greater than that in the E Region to either side that it has acquired the name *auroral electrojet*. During highly disturbed periods the current in the auroral electrojet sometimes exceeds 300,000 amperes. The electrical energy involved is enough to power hundreds of millions of 100-watt light bulbs, should anyone figure out a way to hook them up. The task would be difficult, because the auroral electrojet is several tens of kilometers thick and sometimes several hundred kilometers wide.

The magnetic fields associated with the auroral electrojet currents produce significant variations in the geomagnetic field recorded at magnetic observatories located near the auroral zone. The direction of the geomagnetic field can alter by a degree or so, and the field's strength can change by as much as 10 percent. Because they fluctuate rapidly, the auroral zone ionospheric currents induce current in the conducting earth below, and in any long conductors near the earth's surface, such as long telephone lines, electrical transmission lines and pipelines.

The induction produces generally harmful effects in manmade objects. While the currents created in electrical interties are small compared to the 50- or 60-cycle current normally carried, they cause unwanted imbalances in system transformers that lead to long-term damage, bothersome audio noise, tripping of protective relays, and even catastrophic failure. An expensive 735-kilovolt transformer at Hydro-Quebec's facility on James Bay, Canada, failed during a great red auroral display in 1980. Four months later, another major auroral display occurred and destroyed the transformer's replacement. More recently, in March 1989, a huge magnetic storm caused Hydro-Quebec's distribution system to collapse. Six million Canadians lost electrical service, some for as much as nine hours. The magnetic storm caused other distribution and equipment failures as well, including the failure of a transformer at one nuclear generation plant.

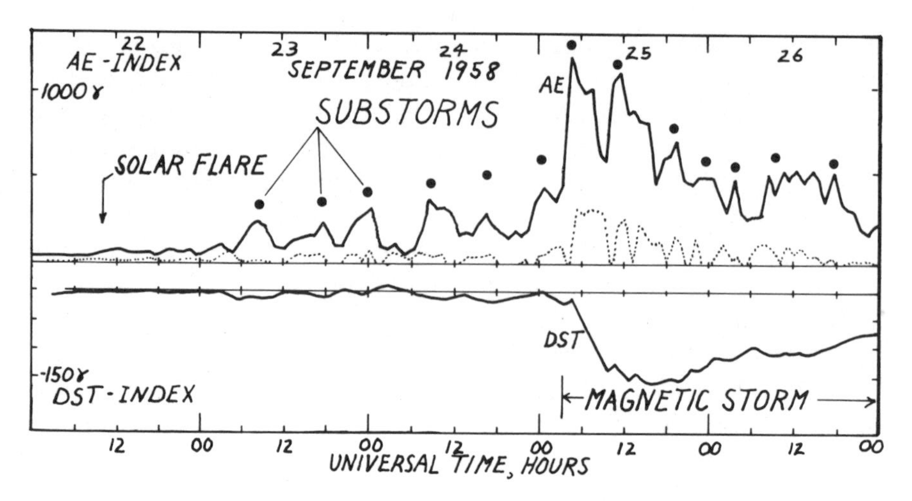

Figure 13.2 Top curve: The AE index. Solid dots along it mark the occurrence of auroral substorms. Bottom curve: The Dst index. Its bay-like excursion early on September 25, 1958, marks the beginning of a magnetic storm. The dotted curve at center indicates bursts of energy injected into the Van Allen belt.

Also expensive is the corrosion the induced current produces in the trans-Alaska oil pipeline. Indirect measurements indicate that currents near 1000 amperes sometimes flow in the pipeline, and as these leak out through buried portions of the pipe they speed up its disintegration.

A useful indicator of auroral activity—and indeed of the substorm activity that affects the entire magnetosphere—is an index based on the aurorally associated ionospheric currents. Called the AE index (A for auroral, and E for electrojet), the index is compiled from magnetic data collected by permanent magnetometers spaced around the northern auroral zone. The index well depicts the burst-like substorm behavior and serves as a qualitative measure of the energy injected into the atmosphere and into the Van Allen belt during substorms. Another magnetic index, Dst (for disturbance, storm time), derived from magnetic observations at low and middle latitudes, provides a quantitative measure of the energy content of the Van Allen belt region. Together, the AE and Dst indices portray the general condition of the magnetosphere and its instantaneous levels of activity. An example of the two indices on the same plot appears in Figure 13.2. Here the lower (Dst) curve shows the worldwide depression of the geomagnetic field at middle and low latitudes during a magnetic storm. The top curve, a plot of the AE index, shows the burst-like nature of substorms, each one marked with a solid dot. The middle curve represents energy injected into the Van Allen belt during the substorms.

Section 13.3 OTHER AURORALLY RELATED PHENOMENA

During periods of auroral disturbance—or to put it more specifically, during auroral substorms—an increased flow of energy passes inward through the outer magnetosphere, some entering the Van Allen belt trapping region within the inner magnetosphere, and the rest skirting poleward of that region to dissipate in the high-latitude atmosphere. The energy transfer involves a variety of processes that may convert energy from one form to another. As particles cross the outer boundary of the magnetosphere, some of their kinetic energy (energy of motion) alters the geomagnetic field and in that process some of the kinetic energy may become stored in the field.

Just as it is possible to assign a specific amount of energy to a moving particle (the concept of kinetic energy), it is possible to assign given amounts of energy to magnetic and electric fields. If a magnetic or electric field occupies a region, the field locks up an amount of energy proportional to the square of the strength of the field. Scientists speak of magnetic energy density or electric energy density, and if they want to know how much energy the field contains they integrate the energy density over the volume of space occupied by the field. By this means, they can discuss the interchange of energy between its various forms. The aurora and its related phenomena involve interchanges among kinetic energy, magnetic energy, electric energy and also energy stored in the structures of atoms and molecules. When that latter form of energy converts to light energy, the result is the visible aurora.

The key processes involved in the auroral substorm are not yet all understood. Some evidence suggests that the substorm merely represents an increased flow of energy into the magnetosphere from the solar wind. However, the burst-like nature of substorms and some of their other characteristics suggest that the substorm might be a process internal to the magnetosphere. Kinetic energy from the solar wind may be converting to magnetic energy stored in the outer magnetosphere, and the substorm may be a sudden conversion of this energy back into kinetic energy that particles carry into the auroral atmosphere and the Van Allen belt trapping region. Substorms may be the outer magnetosphere's belches, although another phrasing might be more apt because the energy passes on through the region to the atmosphere and the inner magnetosphere. The energy transfer certainly is not steady-state; it involves many transient processes with time scales ranging from a tiny fraction of a second to hours.

Transient processes typically generate turbulence and waves. A rock dropped in a pond generates waves that travel across the water, and the plucking of a violin string produces audio waves. During substorms, a certain amount of rock-dropping and string-plucking transpires in the magnetosphere and the ionosphere, generating waves of various sorts. Some are radio waves that travel with the speed of light; others are of lower frequency, with periods up to minutes and which travel slowly through the system. Some of these waves move along geomagnetic field lines and others across them, creating periodic variations in the observed geomagnetic field. The behavior is as though the geomagnetic field lines were being plucked by some giant unseen hand. The various waves running around in the magnetosphere and the ionosphere may be important to the particles that are moving through too, because the particles and the waves may interact in ways that might slow down or speed up the particles. Wave-particle interactions—the waves speeding up or slowing down the particles, and the particles modifying the waves—perhaps are the cause of the pulsating aurora.

Observations show that during substorms, in addition to the generation of an almost bewildering array of waves, other transient processes occur. Increased fluxes of X-rays penetrate into the auroral atmosphere down to altitudes of 30 km, low enough to be detected by balloon-borne instruments. Modification of the magnetosphere during storm periods affects its ability to shield out cosmic rays so that variations in their fluxes occur. The impact of incoming auroral particles and the electric fields associated with them and the convection of the outer magnetosphere heat up the high atmosphere and impose motions on it. In the high atmosphere of the polar and auroral zone regions, the speed of winds (the motion of the neutral atmosphere) sometimes is great. Above altitude 100 km, the wind speed exceeds 150 km per hour nearly every day, and during disturbed times it becomes greater than 300 km per hour. The high-altitude wind at the auroral zone is typically to the west in late afternoon and to the east in early morning. These directions fit with the cause of the wind: the drag imposed on the high atmosphere by the magnetospheric convection pattern.

Section 14
FOLKLORE AND LEGENDS

Section 14.1
NAMES FOR THE AURORA

Some of the names applied to the aurora during historic time give hints about how people interpreted the auroras they saw, and some of the names also hint at the folklore and legends. Ancient Greek and Roman names included: *chasmata* (chasm), *hastea* (javelin), *abysses* (deep pit), *trabes* (beams), *bolides* (darts), *faces* (torch), *saggitta* (arrow) and *pluvia sanguinea* (blood rain). Ancient Chinese names were *Chu-Long* (candle dragon), *thien lieh* (cracks in heaven) and *thien chiens* (swords in heaven). English-language names include flying fires, flying dragons, fire drakes, merry dancers, northern streamers, burning spears, Morris-dancers, pretty dancers, marionettes and Buddha's lights.

Some literal translations of former or current names for the aurora used by Europeans and Asians include: glowing snakes, bloody weapons, angry spirits, bloody armies, fire signs, north shine, north stream, heavenly fight, wind light, fox fire, sky fires, dancing goat, flying dragon, fiery heavens and battle lines of fire. Translations of some Eskimo and American Indian names are: ball player, sky dwellers, that which moves rapidly, untimely birth, caribou cow, the old man, and dance of the dead.

Galileo apparently was the first to apply the name aurora to the northern lights. In 1619 he named it after Aurora, the Roman goddess of morning who appeared as a forerunner of the sun prior to the start of each new day. Galileo mistakenly thought that the auroras he saw were due to sunlight reflecting from the atmosphere, and he used his misconception as an argument against Roman Church doctrine that the earth was the center of the universe. Ordered by the Roman Inquisition not to write on astronomical matters, Galileo was then writing under the name of one of his students, Mario Guiducci.

In North America the names northern lights and aurora are used interchangeably, but aurora is becoming more common. The name *aurora borealis* also is used for the aurora of the northern hemisphere, and *aurora australis* for that in the southern. Even so, most Australians, owing to their northern heritage, call what they see the northern lights. Occasionally people use the term *aurora polaris* to mean the aurora of both northern and southern hemispheres, but aurora suffices.

Section 14.2 ESKIMO LEGENDS OF THE AURORA

Eskimos have inhabited the region of frequent auroras that extends over the region from western Alaska to Greenland for at least 6,000 years. Just as the Eskimo language varies from one village to the next across this great expanse, the legends about the aurora differ from place to place, and perhaps each area had more than one legend. Some of the legends are very old, having been passed down to the present with care. Emily Brown, an Eskimo lady who collected many legends from her relatives in western Alaska, said that storytellers practiced for celebrations carefully. If a story was not told right at the final dress rehearsal for the celebration it would be deleted from the program, perhaps to be practiced more for telling at a later date. The elders also told the youngsters the same stories over and over at bedtime. Brown hinted at an evolution of some of the stories through a sharing of versions when storytellers from different villages got together, so the legends have not always stayed the same.

According to Brown, a good Eskimo legend should challenge the listener's imagination. Much Eskimo legend about the aurora and other natural phenomena deals with spirits. In the legends the spirits are everywhere: intertwined in the now and the hereafter; in all the animals and fishes; and in the land, the sea and the sky. The teller of a legend and his listener perhaps can be complacent about the spirits since they are beings remote from human affairs. On the other hand, perhaps a person best take the spirits seriously because they do indeed hold awesome powers.

That ambiguity in attitude was illustrated by the behavior of a group of Eskimo high school students in Nome, Alaska, some years ago when I gave a lecture on the aurora and mentioned the well-known superstition that a person should not whistle at the aurora for fear it will come down and cut off the whistler's head. All in the audience smiled, but many nervously giggled and hunkered down in their seats. "We don't believe it," one boy spoke out, "but then we are not going to go out and whistle at the aurora either." Laughter broke out, some covering continuing nervousness, but most in appreciation of this boy's clever Eskimo-style joke.

Many Eskimo legends are more subtle than some writers reporting on them have believed. Emily Brown noted that many of the legends actually were composed primarily as media for teaching moral philosophy and for instilling proper behavior in children. Thus many are like the fables and fairy tales told to children of other cultures. That some of the legends might be misinterpreted by non-Eskimos was suggested by Emily Brown one day when queried about something an anthropologist had cited. "Oh, those silly old anthropologists never get things right," she said, "You can tell them anything, and they will believe it."

Some of the reported superstitions and legends, leaving out most of the lessons in moral philosophy, follow.

A widely held Eskimo legend: A departed spirit may enter the body of a new-born child, so the child's parents must be cautious in their treatment of the child; too much punishment may cause the spirit to leave. Departed spirits also go to various levels of afterlife; the level achieved depends on a person's behavior in life and the manner of death. A person who dies of sickness or other routine cause and who has not been a good person usually ends up in a bad place, perhaps beneath the sea or in the bowels of the earth. Punishment is not the object, but this place may be very dark and stormy, and much snow and ice covers the area. Another possibility is a sort of intermediate level of the hereafter that is monotonous but free of cold and hardship. Best of all is hereafter in the aurora, the highest level of heaven. This is a happy place where no snow falls, nor is it ever stormy. Within the aurora, brightness abounds, and many easily caught animals live there. The person who goes to this highest level of heaven in the aurora is the man who dies in the hunt, the woman who dies in childbirth, the person who commits suicide or is murdered. It helps if this person has always been generous to the poor and starving.

Another widespread Eskimo legend: The aurora is a game of football, a game that in real life is played by groups of men on a strip of beach perhaps several miles long and which can take more than a day to complete. Any tactic is fair, and when one team gets the ball to the end of the beach that team wins. In some legends the aurora is the spirits playing football with a walrus head, and the contra-streaming movements of the lights across the sky and any accompanying noise are the evidences of the struggles among the spirits. But on Nunivak Island, off the west coast of Alaska, the walrus get their revenge: the aurora is the spirits of walrus playing football with a human skull. In parts of Greenland the aurora is believed to be caused by the spirits of stillborn or murdered newborn children

playing ball with their afterbirths, and while doing so they may whistle and hiss. A legend from Norton Sound (Alaska) says that the spirits play their football game with the heads of disobedient children. These hissing, biting spirits like to entice up into their midst other children who do not mind their parents.

A legend of northern Canada tells that the aurora is the spirits of the dead dancing in an ethereal light to entertain themselves when the sun is absent. They wear multi-hued clothing. And over the land and sea of the Hudson Bay area the sky forms a dome of hard material arched high overhead. A dangerous path leads up to a hole through which only the raven, suicides and victims of violent death have passed, and they light torches to guide new arrivals. In that auroral light it is possible to see the spirits feasting and playing football with a walrus skull. The feet of the spirits sometimes make a whistling, rustling, cracking noise as they dash across the frosty snow of the heavens above Greenland and Canada. In the olden days, the auroras over western Alaska did not merely whistle and rustle, they howled.

If a person whistles with a sound such as the aurora makes, it is sometimes possible to communicate with the dead. Proper commands, like whistling and the right sort of hand-rubbing, can make the aurora dance closer or skip farther away. This is a widespread legend in North America. In western Canada, spitting at the aurora can make the various forms of light run together and change shape. In the vicinity of Point Barrow, Alaska, the aurora is to be feared, and it may be wise to carry a knife for protection from the lights. Care should be taken not to offend the aurora, but it can be scared away by throwing frozen dog excrement or urine into the sky.

A man of the Diomede Islands once received help from the aurora when he was a little boy. The aurora told him the best way to conduct his life, and he grew up to be his village's leader. Those named after him also received help from the aurora, but after five generations the power was almost gone.

Section 14.3 OTHER BELIEFS AND LEGENDS

The aurora as fire figures often in the legends and myths of North American Indians. The Mandans of North Dakota thought the aurora was the fire of northern warriors cooking their enemies in large pots. Washington's Maka Indians said the auroras were fires lit by a powerful tribe of dwarfs in the far north, while the Menomini Indians of Minnesota saw the aurora as the light of torches used by friendly giants to help spear fish at night. To the Ottawa Indians, the aurora was a great fire lit by their creator to remind them of his interest in their welfare. The Fox Indians of Wisconsin considered the aurora to be a bad omen, a gathering of ghosts of slain enemies eager for revenge. The Eyaks and Tlingit of Alaska believed the aurora to be the sign of an approaching battle that would bring death.

Indians living in the Koyukuk region of northwestern Alaska saw the aurora as a friendly thing that could be attracted by banging on metal pans. The Chippewa Indians of central Canada said that the appearance of aurora was good because it meant that many deer were in the sky. One writer has suggested that this legend may have devolved from the observation that the stroking of deer hair leads to electrical discharge that, like the aurora, can be seen at night.

Some North American tribes thought of the aurora as "the old man" dressed in glowing robes about which his long white hair streamed and who flashed his fiery eyes. The Creek Indians of Georgia and Alabama and the Cheyenne Indians of Wyoming believed that the occurrence of the aurora meant the weather would change. The Penobscot Indians of Maine said the aurora would bring a windy day. If the aurora flickers, the wind will be strong and steady, and if the aurora is steady and quiet, the wind will be gusty.

Europeans and Asians also have their legends. In Lapland, as in some parts of the North American Eskimo word, people should be cautious about the aurora. If they mock the lights, the aurora might kill them with an axe or burn them. To whistle at the aurora is still dangerous today in Lapland because it will come closer and perhaps tear out one's eyes. Sleigh bells should not be used when the aurora is out, nor should women go bareheaded because the aurora might seize their hair.

Another Lapp view is that the aurora is but a wintertime version of summer's thunderstorms, and useful for night travel across the snow. Legends of the Russian Lapps portray the aurora as fearful battles between the spirits of those who have been murdered. They stab each other, and cover the floor of the house of the dead with blood. Yet another Russian Lapp legend explains the red parts of the aurora as the blood spilled when the sun's rays accidentally strike the body of a mysterious male figure in the sky, causing his death. The aurora is associated with the heaven god to the Chuvash people of Siberia. He helps women through the agonies of childbirth, and when the aurora is bright it is giving birth to a son.

In Scandinavia and other parts of Europe, people have long considered widespread auroral displays to be the portents of things to come: changes in weather, the deaths of kings and the coming of terrible wars or pestilence. In parts of Norway the aurora was thought to be the reflections of light from schools of herring in distant oceans, the reflections of light from Iceland's geysers and from icebergs in the North Atlantic. Old Finnish and Estonian myths suggest that the aurora is caused by a whale or other monster in the oceans. When it strikes the water with its tail, great splashes of aurora fill the skies. One Finnish legend says that the aurora is fox-fire, the light that comes when the fox turns and flashes his tail, but no heat comes from the fire.

Widespread as they are, auroral legends and myths have sometimes been insufficient to avoid the great terror caused from time to time when brilliant displays have spread over the skies of Europe and North America. The great displays, especially those occurring after a period of prolonged auroral quiet, signaled to many the immediate end of the world. More than once during the past few centuries, Europe's and Colonial America's churches suddenly filled with those preparing to meet their Maker as the lights dashed overhead. Even within the past half-century, that interpretation was not uncommon.[1]

---

1. In 1941 my family was staying briefly in a small coal-mining town in West Virginia; there I saw my first aurora, a brilliant red one that filled the sky. Most of the town's residents believed that the end was nigh so they ran out into the streets where they entered into group prayer, religious songs and a public recanting of past sins. The most memorable part of it all was the confession of a neighbor lady who apologized at the top of her voice for what to my youthful ears were some highly interesting social activities. She remained in her house for several days afterwards.

Part III
AURORAL MYSTERIES

Section 15
UNKNOWNS ABOUT AURORAL PROCESSES

Amazingly enough, auroral scientists still do not understand why aurora occurs in discrete auroral forms instead of as a diffuse glow across a broad region of the sky. The reason somehow has to be bound up in the way matter in the plasma state behaves—it seems to abhor uniformity. We are generally familiar with the way matter in the solid state behaves, and we know that when a solid is heated it converts to the liquid state. Further heating turns the liquid to a gas, and if a gas is heated it becomes ionized and then is said to be in the plasma state. The study of the aurora is a study in plasma physics because the entire region of concern—all the way from the center of the sun to the bottom of the ionosphere—is in the plasma state. In fact most of the universe is in the plasma state; familiar exceptions are the outer half of the solid earth, the oceans and the atmosphere above. The lower ionosphere is called a lightly ionized plasma, for here most of the particles are electrically neutral. The high ionosphere, the magnetosphere and the solar wind are fully ionized plasmas.

Much is yet to be learned about plasmas, and one of the reasons for studying the auroral processes is that they are taking place in a giant natural plasma laboratory close at hand. What is learned there can also apply to processes taking place in and on the sun, in the stars and in the plasmas of interplanetary and interstellar space. The aurora can light our way to a more complete understanding of the entire universe.

Coming back to matters of less grand scale, it is fair to

state that scientists still lack is a full comprehension of how and exactly where particles contained in the solar wind enter the magnetosphere boundary. That problem fringes on another, the nature of the auroral substorm. The substorm may be simply the expression of changes in the rate of entry of particles and energy into the magnetosphere, or it may be at least in part an internal reaction of the magnetosphere that involves conversions of energy from one form to another.

Nor is it known in detail how charged particles, once inside the magnetosphere, become accelerated enough to penetrate as deeply as they do into the polar atmospheres. Several processes may be operating, both far out in the magnetosphere's equatorial plane and lower down in the region a few thousand kilometers just above the polar regions, where in fact acceleration processes have been identified.

Lacking also is knowledge of the relative importance of the interactions that take place in the atmosphere to generate auroral light. Most or all of the processes are known, but how important each is remains uncertain. Direct excitation by particle impact is a certainty, but some observations suggest that other processes may also be operating. These may involve heating by electric fields and interactions between ionized atmospheric constituents and various types of electromagnetic waves that pervade the medium. Questions still remain about all the processes that go into producing two of the aurora's brightest and most simple emissions, the oxygen red and green lines at 6300 A and 5577 A, and the cause of the 'enhanced aurora' remains a mystery.

In short, unknowns abound throughout the vast region that begins on the sun and ends in the polar atmosphere. Nevertheless, scientists have made giant strides during the past few decades, and it is not amiss to claim a general understanding of the aurora.

Many techniques are now in use to gain additional information about the aurora and the physical processes causing it and related phenomena. Some of the procedures and tools used to obtain information routinely:

1. Regular observation of the auroral zone sky with all-sky auroral photographic or highly sensitive video cameras that take frequent pictures during the dark hours.

2. Permanent observatories that use magnetometers to monitor magnetic variations at many locations along the auroral zones and elsewhere around the world.

3. Measurement of the voltages induced in the earth's surface by ionospheric currents, using 'earth-current' recorders, a method that gives little quantitative information but is highly useful for estimating auroral activity when the sky cannot be seen, as during the day or when the sky is cloudy.

4. Regular recording of the behavior of the ionospheric layers by means of ionosondes, devices that send radio waves into the ionosphere and receive the signals reflected back.

5. Systematic observation at the earth's surface of electromagnetic waves generated naturally in the ionosphere and the magnetosphere, and monitoring the propagation of man-made radio signals through the auroral regions.

6. Routine observations of auroral emissions using recording spectrographs that detect how much of each color of light is emitted.

7. By means of receiving devices called riometers, regular measurement of the auroral ionosphere's attenuation of radio waves coming from stellar sources mostly within the Milky Way.

In addition to the routine observations are those of more complex nature made in pursuit of the solution to identified problems or in conjunction with experiments performed using rockets and satellites, as for example, barium releases. The observations might be made from permanent or temporary ground stations, and some even from airplanes.

1. Observation of the sky with photographic cameras or with sensitive television systems which can record even subvisual phenomena and the locations and rapid variations of natural or man-made auroras and chemical release clouds.

2. Measurement of the light emitted at specific wavelengths by means of photometers equipped with filters that admit light of only certain colors. Some of these devices are so sophisticated that they can measure the emissions in a band less than one Angstrom unit wide. This capability allows determining temperature in the high atmosphere and provides detailed information on excitation processes.

3. Probing or heating of the ionosphere with highly directional, high-power radio transmitters in ways that allow measurement of the temperatures and densities of atmospheric constituents, and which may also generate artificial electromagnetic waves in the ionosphere that can be observed to give information on the ionosphere.

4. Observation of the characteristics of the high atmosphere using powerful laser beams.

5. Operation of extremely specialized photometric and spectroscopic devices to study the auroral emissions or those produced by chemical release clouds.

High-altitude balloons are used to carry detectors that measure incoming fluxes of X-rays associated with auroras and to measure electric fields in the atmosphere below auroras. Balloons are useful because they are relatively inexpensive, can be flown anywhere, and can maintain instruments at altitudes as high as 30 km for many hours.

Rockets, satellites and deep-space probes have proven highly valuable for auroral studies and examination of the whole region between the aurora and the sun. They can carry instruments that measure the densities, speeds and directions of motion of electrons, protons and all other species of particles resident in the sun-earth region. The rockets and satellites also permit the deployment of almost every conceivable instrument into and beyond the aurora, including those that would be useless on the ground because of the obscuring effect of the intervening earth's atmosphere. Satellites, however, cannot fly through the lower parts of the aurora because the air is too dense there. Also, satellites fly so fast that they make difficult the detailed measurements of small-scale structures such as the incoming particle streams above individual auroral forms. Rockets are far better for this purpose, and they allow in-situ measurements in the whole height range from the ground up.

In addition to allowing in-situ measurements of incoming charged particles and atmospheric constituents, rockets and satellites can carry devices to measure magnetic and electric fields, radio wave emissions and waves of other types. Best of all, they allow us to see the aurora from above and to remotely sense auroral phenomena by techniques ranging from direct photography aboard manned satellites to the use of fully automatic imagers (like the DMSP imagers) that routinely acquire pictures of auroras and clouds from locations high above the earth's surface.

Section 16

AURORA AND WEATHER

Their anecdotal experiences of seeing aurora have led some northerners to believe that a causal relationship exists between low outdoor temperature and the occurrence of aurora. However, the relationship noted is simply because, as Plate 26 shows, it is possible to see aurora only when the sky is clear, and when the sky is clear, outdoor temperatures are typically lower than when the sky is cloudy. The observation is correct, but the conclusion wrong. As yet, scientists have not found conclusive proof of any direct connection between weather and the appearance of aurora and related solar-terrestrial phenomena. On the other hand, a few studies and observations do hint at a connection.

The question is important because we humans want to know everything possible about what creates changes in the medium we breathe and live in. Our supply of food and water is critical, and we know that they depend on weather and climate. Other concerns such as ozone depletion and global warming seem almost as serious.

That climate change has occurred in the past is certain, but how that change has been brought about is problematical. The most obvious and likely fundamental cause is variation in the solar output of light energy, the energy that arrives at the earth 8 minutes after leaving the sun. This energy is of order 100 times greater than the solar energy that streams out in the solar wind to arrive at the earth a day or two later and then produces the aurora and its related phenomena. Therefore, a tiny change in the output of solar light can have a far greater terrestrial effect than can a proportionally larger change in the energy carried by the solar wind.

A tantalizing hint of intermediate-term change in the sun's output of energy and of its relationship to observed solar variations comes from the Maunder Minimum. This was a time of virtually no sunspots (see Section 6.3) and of prolonged cold in North America and Europe. During the intervening 250 years, scientists have made regular observations of sunspot activity and have documented the 11-year (approximate period only) solar cycle and variations in the lengths and strengths of the individual cycles. When it became possible to measure indirectly the solar energy arriving at the earth's atmosphere, the observations did not indicate variation greater than the estimated observational errors. However, the advent of satellites in recent decades has allowed more accurate measurements of the solar output at locations outside the earth's atmosphere. Evidence is accumulating that the solar output does indeed decrease during periods of sunspot minimum relative to that observed during sunspot maximum.

If continuing measurements bear out the apparent relationship between solar light output and sunspot activity, then proof of a direct connection between aurora and weather may be difficult to determine. That is so simply because the aurora statistically varies in concert with the sunspot variation, and so what might look like a cause-and- effect relation could be merely coincidental. One recourse is to look for possible short-term connections between aurorally related phenomena and weather patterns.

One study conducted some years ago gave at least minor evidence that storms initiated in the North Pacific during magnetic storms (times of major auroral displays) tended to grow more than those developing at other times. Substantiation of this result is required before awarding it too much credence; however, the idea that auroral phenomena might have some triggering effect on storm development is reasonable. Certainly it is known that the aurorally related phenomena do heat the high atmosphere and enhance its circulation in the polar regions. The energy involved is relatively minor compared to that involved in primary weather processes, so nothing more than a tickling of other processes seems possible. But a tickling of a dog's foot can lead to frantic activity, and the atmosphere's weather-making processes might behave likewise. One mechanism that might do the tickling is the ionospheric electric field associated with auroral displays. Cloud growth is highly sensitive to the strength of the electric field above thunderstorms, and so it has been suggested that ionospheric electric fields may affect the electric field lower down enough to modify storm development. Of possible relevance are recent studies showing that lightning strokes sometimes extend from cloud tops to the ionosphere.

Section 17
THE MYSTERY OF AURORAL SOUND

Many people have reported hearing sound in conjunction with seeing aurora, but no audio recording or instrumental observation to date has verified that sound is associated with the aurora. The mystery of auroral sound is an intriguing one because it poses questions both about physical mechanisms acting in the atmosphere and about human and animal sensory systems.

Persons who have reported hearing auroral sound have sometimes done so sheepishly or with a defensive tone of voice because they expected the audience to be skeptical. In some sense or another, however, aurorally associated sound probably is real. Yet, the only observations are anecdotal, and that is not satisfactory proof. Any phenomenon reported only by anecdotal means deserves a skeptical audience, and auroral sound is no exception. Only when undeniable evidence is available, perhaps in the form of repeated instrumental recordings, are we likely to be convinced completely of auroral sound's reality. Too few attempts have been made to record auroral sound with instruments, so the fact that no instrumental recording exists does not provide a satisfying refutation that the sound is real. Furthermore, the possibility exists that a person can perceive 'sound' when no actual sound waves impinge on the person's ears. For example, some people with a variety of metal dental fillings can 'hear' broadcast band signals when they are located close to a powerful transmitter. The fillings rectify the strong electromagnetic signals and convert them to vibrations in the audio range. No sound actually enters the ear; nevertheless, the person hears the program that is broadcast.

This example points out another aspect of the problem of auroral sound. The reports of hearing broadcast stations without benefit of a radio receiver are—like those of hearing auroral sound—anecdotal. Yet scientists are inclined not to doubt the reports because an easy explanation is at hand: the rectification of electromagnetic signals by teeth fillings. Furthermore, a subject claiming to hear radio stations can be tested simply by placing him near a transmitter to see if he can repeat the words spoken by an announcer to whom another person is listening through a radio receiver. But when it comes to auroral sound, no physical mechanism is obvious, nor is it possible to test the hearer by again exposing him to a controllable source since no earthly hand controls the aurora. The lack of a known mechanism and the inability to perform adequate testing rightly makes any person who likes to deal with logic somewhat skeptical—even some who have experienced the sensation of aurorally associated sound.

Some 300 documented reports of hearing auroral sound are available. Two hundred of them are contained in a paper by S. M. Silverman and T. F. Tuan published in the scientific journal *Advances in Geophysics* in 1973. Another 100 reports are available in a collection assembled from documents collected over the years by personnel of the University of Alaska's Geophysical Institute. These documents consist of solicited and unsolicited letters from individuals, completed questionnaires, informal reports and write-ups of oral reports. Nearly all of the reports in the two collections were written or orally presented well after the times when the auroral sounds were experienced, sometimes years or decades later. Hence nearly all suffer from a lack of specific information about the dates or times of the reported incidents, and vagaries of memory or subsequent experience may have influenced many of the descriptions contained. Somewhat countering these probable influences is the evident fact that many hearers of auroral sound thought that their auditory experiences were unusual and highly memorable events. Whatever details were noticed at the time seem to have been well remembered.

Section 17.1
SELECTED EXTRACTS FROM OR SUMMARIES OF REPORTS ON HEARING AURORAL SOUND

Early-day explorer David Thompson wrote of an incident during the winter of 1796-97 at Reindeer Lake, Saskatchewan:

> In the rapid motions of the Aurora, we were all persuaded that we heard them, reason told me that I did not but it was cool reason against sense. My men were so positive that they did hear the motions of the Aurora, this was the eye deceiving the ear: I had my men blindfolded by turns and then enquired of them if they still heard the motions of the Aurora. They soon became sensible that they did not and yet so powerful was the illusion of the eye on the ear that they still believed that they heard the Aurora.

Yukon explorer and surveyor William Ogilvie, writing of an incident on November 18, 1882:

> ....I often listened [under favorable conditions, and] I cannot say that I ever fancied I heard anything. I have often met people who said they could hear a slight rustling noise whenever the Aurora made a sudden rush. One man, a member of my party...was so positive of this that...when there was an unusually brilliant and extensive display, I took him beyond all noise of the camp, blindfolded him and told him to let me know when he heard anything, while I watched the play of the streamers. At nearly every brilliant rush of the auroral light, he exclaimed: 'Don't you hear it?' All the time I was unconscious of any sensation of sound.

Hans Jelstup, Norwegian astronomer, describing an incident on a hilltop near Oslo on October 15, 1926, while making determinations of longitude:

> I was at work in a field observatory with a transit instrument registering star transits and chronometer beats for time determinations, when an initial aurora attracted my attention...I was able, during intervals between my observations...to observe the aurora, which was certainly one of the most splendid I had ever seen. But what is of predominant interest is the following fact: when, with my assistant, at 19h 15m...I went out of the observatory to observe the aurora, the latter seemed to be at its maximum: Yellow-green and fan-shaped, it undulated above from zenith downwards—and at the same time both of us noticed a very curious faint whistling sound distinctly undulatory, which seemed to follow exactly the vibrations of the aurora.
>
> The sound was first noticed by me, and upon asking my assistant if he could hear anything, he answered that he noticed a curious increasing and decreasing whistling sound. We heard the sound during the ten minutes we were able to stay outside the observatory, before continuing our observations. [A few minutes later they again went outside and both the aurora and sound had vanished. During this observation, auroral scientist Carl Störmer, located elsewhere but in the general vicinity, was making determinations of auroral heights; the results were that the aurora was of normal altitude, 90 to 132 km.]

Carl Störmer, reporting on an extremely large auroral display seen all over Europe on the night of January 25-26, 1938, while making height determinations of auroras with his triangulation network in Norway:

> Big red draperies and rays, whose height reached towards 700 km had for a time been moving from

> the south towards the zenith and spread now all over the sky as an imposing corona with white, blue, yellow green and red rays. It was the most imposing display I had ever seen [Störmer had seen many.] It is worth mentioning that during this corona, one of my auroral stations on Njuke mountain in Tuddal, 733 m. above sea level, more than 100 km west of Oslo heard a sound apparently connected with the aurora.
>
> My chief assistant...on that station describes the phenomenon as follows: "During the imposing display of the big corona when the whole sky was filled by flaming rays my assistant and myself heard a strange sound coming from above, first from SW, then from Zenith, and last from NE. The sound lasted about 10 minutes. It rose to a maximum and faded successively as the intensity of the aurora diminished. I had the impression it had something to do with the white rays. My assistant first informed me about the sound and I immediately took off the microphone from my ears and went some steps aside to hear better. In fact, with the microphone on the ears I did not hear anything; thus the sound could not come from the telephone." [The measured altitude of this aurora was 95 km. Another person located in the valley below the auroral observing station also reported hearing the sound.]

Charles Wilson of the Geophysical Institute, a leading authority on auroral infrasound, reported orally on an incident in July 1958 when he was the auroral observer at Little America. He had climbed atop the auroral tower at the station to check the operation of the auroral all-sky camera when, although not trying to listen for sound he heard a sound that "was not at all subtle—not loud but distinct." The sound was a whooshing noise that repeated several times during an interval of some tens of seconds. Several of the whooshes correlated with surges of ray motions along closely spaced auroral arcs overhead. The rays were tall and the entire sky was covered with red aurora. Despite having lived for many years near Fairbanks and having often observed the aurora, this was the only time Wilson ever heard aurorally associated sound.

Owen Robbins, a weather observer on Alaska's North Slope in 1979, wrote that he had heard aurora on several occasions. Describing one, he said:

> At approximately 11:40 pm I went outside and before looking up I knew the aurora was overhead for I heard it. Whenever I first hear it I am prompted to duck, so as to avoid being hit on the head. I looked up and it was overhead, running an east-west pattern across and [filling] the sky. I remained outside from then on, only returning to record my weather and to get my tape recorder [which did not work because of low batteries. He described the sound as "a distant crackling hiss—like an object zipping over one's head. ssssssSSSSsssss. Not unlike a spitting cat."].
>
> When the display is extraordinarily active the aurora has a habit of curling up on itself [the description of a spiral or surge forming]....Also, when the Northern Lights are going about like this, the sound is most prominent; additionally the sound is audible, yet only slightly, when the motion is rapid in the extended bands. I have never heard auroral sound when there was an absence of either a curling motion or extremely rapid movement. [He said he could not hear the aurora through his parka hood, but heard it with no varying degree of clarity whether standing, sitting or lying down, positions I suggested that he take if he heard the sound again after first reporting it. Mr. Robbins said he had partial hearing loss, 60 percent in one ear and 10 percent in the other.]

Dr. Carl Benson, glaciologist, years afterward reported hearing aurora during September 1950 while with a geologic party in the foothills of the Brooks Range where little vegetation other than low ground cover was present:

> [On this particular night a very bright aurora appeared overhead; the sky was clear and there was no wind or noise from any other source.] I had the impression of a fast swishing and crackling, the crackling being subdued relative to the swishing. Overall, the sound lasted five or ten minutes, but was strongest during an interval of about one minute. The sound was variable; it came and went with the rapid motion of the aurora. [Benson said the aurora was brilliant with multicolored hues. One other geologist present thought he might be hearing the sound, but two others present heard nothing. Benson offered a possible explanation by stating that, in earlier years, he had excellent hearing and eyesight, and when out in the field he would always be the first in the party to hear an approaching aircraft. He also said that, in retrospect, he wondered if he really had heard the aurora or just thought he was hearing it because this was the first really good display he had ever seen and perhaps was unduly impressed by it.]

Paul Solka, Jr., writer and artist, from an article published in the *Fairbanks Daily News-Miner* on February 6, 1970:

> [The described incident occurred in mid-December 1919 when Solka was young and living with his parents at Richardson, a bygone settlement 70 miles southeast of Fairbanks. The family was inside a cabin on a cold, calm night when one of two sled dogs outside began to growl and bark as though someone were approaching. The family went outside.] The sky was bright with dancing aurora. They moved with incredible speed across the sky, then seemed to hang like transparent drapes between us and the nearby hills....Some of the streamers seemed to dart within arm's length and then

> fade away, only to be replaced by others. As they moved, a soft hissing sound accompanied them. This was what had disturbed [the dog] and his odd behavior was perhaps the only reason we saw and heard this rare display. [Solka described the sound as like that of a hand rubbed lightly over several sheets of crumpled tissue paper. He stated that he and his family had listened for the aurora many times during the 50 years since that incident, but had never heard the aurora again.]

John R. Craig (Silverman and Tuan, Report No. 46):

> About the 15th August, 1882...I [with three others] camped on the prairie some 20 miles east of Fort McLeod [Canada]. We built our camp fire and had supper and soon after retired to rest. The night was calm and bright. Lying awake in the tent, I heard a mild crackling noise which brought me outside quickly, fearing that our fire had not been thoroughly extinguished. The fire was dead, but the heavens to the north were showing a greater display than I had ever seen. The aurora was shooting upwards and receding with almost lightning rapidity and with varying colors. A broad yellowish splash of flame spread across from the west to the east, ascending from the horizon and proceeding with what I can best describe as a swishing noise, while at the same time a crackling noise accompanied the darting and shooting of the aurora....The sound from the aurora was clear, distinct, impressive and so indelible that the forty years which have elapsed left the audibility of this grand display fresh and clear....

Robert R. Racey (Silverman and Tuan Report No. 56), contained in a letter written in 1938:

> During the winter of 1894 and 1895 I [was driving a horse] in the Upper St. Maurice River

> country [Canada]...I drove all night southward....Temperature was probably 30 degrees below zero. The atmosphere appeared to be quite clear, the heavens brilliantly illuminated with stars.
>
> My back was turned to the north, and somewhere below the Rat River I became conscious of much light behind. I then heard prolonged, regular swishing sounds, again and again, which resembled music produced when the strings of a harp are lightly touched. [He describes a spiral with rays.]....and each time a ray shot upward it was accompanied by the "swishing" sound, and only then was the sound audible....

Dr. H. D. Curtis, astronomer in charge of the Labrador Station of the Lick Observatory in 1905 (Silverman and Tuan Report No. 65, also published in *Science*, September 1921):

> The station was located at Cartwright...and auroral displays were frequent and bright during July and August. On several nights I heard faint swishing, crackling sounds, which I could only attribute to the Aurora. There were times when large, faintly luminous patches or 'Curtains' passed rapidly over our camp; these seemed to be close, and not more than a few hundred feet above the ground, though doubtless much higher. The faint hissing and crackling sounds were more in evidence as such luminous patches swept past us....I tried in vain to assign the sounds heard to some reasonable source other than the aurora, but was forced to exclude them as possible sources; besides, what I heard didn't *sound* like anything from anything I could postulate...,In short, I feel certain that the sounds I heard were caused by the aurora and nothing else. There was, moreover, a certain synchronism between the maxima of these sounds and the sweeping of auroral curtains across the sky.

A. J. Woodward of Mimico, Ontario (Silverman and Tuan Report No 127, a report said to be based on many years of daily meteorological observations):

> There is a distinctly audible noise from the aurora, but only from general displays that produce running waves from horizon to zenith in about one second. If these waves are wide, and consequently nearer, a rustling sound is produced, and the narrower the wave, the sharper the noise, almost to a crackling sound.

G. I. MacLean, Gold Commissioner, Yukon Territory (Silverman and Tuan Report No. 139, written March 1931 in response to a questionnaire on auroral sound):

> Certainly where there is an unusual display of Northern Lights and where they are not only very active, but cover the whole sky, there is a peculiar faint noise, so faint that unless there is absolute silence it cannot be heard. Often I have heard it where the blanket was over my face and I did not know "the lights were on." But whether it comes from the Aurora or not I was not able to determine. It sounds like two pieces of paper being softly drawn over one another. This description, of course, I had heard many times before I had seen the lights. But it fits very well; I made many experiments of stopping my ears and then suddenly exposing them, and always the same sound was again heard. I wondered at the time whether it might not be the action of the intense cold on the eardrums, as it was very akin to the singing in one's ears caused by coming down rapidly from a high to a low level. Moreover, the noise is not steady, but seems to wax as the streamers form and re-form into various patterns. On clear cold nights when there was no display of lights, I have tried to distinguish the same sound, but could not do so.

Summary of observations by personnel of the Geophysical Institute while observing auroras:

> At the Fort Churchill auroral observatory in 1968, Larry Sweet of the Geophysical Institute was located inside a large plastic dome occupied with pointing television cameras at the sky in conjunction with rocket firings. A violently moving discrete aurora appeared overhead, one of the more spectacular ever recorded on television. During the event Sweet called out over the intercom, "I hear it," referring to having heard the aurora. Owing to other conversation over the network, his message was ignored, and in the excitement of working to record the excellent aurora and his other activities he too forgot the incident and did not mention it again until later. The TV recordings showed some of the highest velocities of auroral rays (curls) ever observed.
>
> The next year, graduate students Thomas J. Hallinan and John S. Boyd and photographer Russell Beach were similarly occupied with auroral observations at the Fort Yukon auroral observatory. On one night, when a very bright, active rayed arc with red lower border appeared directly overhead, John Boyd heard what he described as "swishing sounds" at least twice. He said they were easily distinguished from identified background noises such as diesel generators operating in the distance. The sounds apparently coincided with auroral forms swirling through the zenith. Walking briskly with his hood over his ears not far away, Tom Hallinan heard nothing, but the incident caused Hallinan to search for auroral sounds on subsequent nights. Several nights later, Hallinan heard faint noises when active auroras were overhead on two occasions, but none when the aurora was not so situated nor during the next day when he sought to listen for

unusual sounds at several places in the area. The next night when Hallinan was inside manning television equipment, Russell Beach called over the intercom to say that he was hearing sound, just as an extremely bright aurora arc moved into the zenith. Hallinan ran outside and perhaps heard aurorally associated noise briefly but was unsure because of other noise caused by people talking and dogs barking. Beach reported that he and another man heard swishing noises on each of five occasions when the aurora "boiled overhead." The other man with him heard the sounds only after removing his parka hood from his head. This was prior to the time Hallinan came outside. Several hours later, Beach "seemed less convinced that he had heard anything."

After returning home to Fairbanks, Hallinan spent several evenings outside searching for auroral sound. He once thought he heard a distinct noise that lasted for several seconds when a bright aurora swept overhead. A co-worker standing beside him also heard the noise but attributed it to wind in the trees.

My own closest brush with auroral sound occurred one night when inside a noisy auroral observatory on Ester Dome, just west of Fairbanks. I was busy photographing the sky with a television camera when such a spectacular aurora appeared overhead that I telephoned my home, several miles away, to suggest that my wife Rosemarie go out and look at it. She did go out, and while watching a bright array of multiple arcs with fast-moving rays just overhead she heard a sound like that of ice-coated weeping willow twigs rustling in a breeze. Initially she thought the sound was due to wind motion in nearby trees, but satisfied herself that this was not the cause, nor was it her parka hood brushing against her hair. She did not notice any correlation between the sound and any particular aspect of the auroral variation. Suspecting that I was skeptical that auroral sound was real, and wishing to avoid any wise-acre remarks from me, Rosemarie did not mention this incident until much later.

The above reports appear here because each includes some information on both the aurora and the sound. Several obvious threads run through these few reports, and also through most of the total collection of 300, only 60 of which give any specific information on the aurora. The most striking is that the sound is reported when brilliant, fast-moving aurora is overhead the observer, and the sound is correlated to at least some extent with the auroral motions.

Approximately 15 percent of all 300 or so reports in the two collections appear to be from scientists, engineers or other technically trained persons, some of whom were actively engaged in auroral observation at the time when the sound was heard. The reports from the technically trained persons differ little from those of others, except that they tend to contain more information on the aurora and the observing conditions when the sound was heard. When afterwards mulling over their observations, some of the scientists became suspicious of their own reports, being well aware of the fallibility of the human observer in unexpected, perhaps exciting situations.

A striking general characteristic of nearly all of the 300 reports, regardless of their sources or of the times of the events, is a consistency in the descriptions of the sounds heard.

The letter 'S' figures heavily in at least 80 percent of all descriptions of auroral sound. Over 90 percent of the descriptions state that the sound is a variable or undulatory hissing, swishing, whooshing or crackling noise: like 'the swishing of a taffeta skirt,' like 'burning grass,' like 'the sound of crumpled tissue paper lightly rubbed between the hands,' like 'the sound of a comb being drawn through a woman's hair,' like 'frying grease,' or like static electricity. Some of the reports stated that the sound was something similar to, but not quite the same as, the description given—that it was a unique noise difficult to describe. The sounds were reported to have durations of a few seconds, a few minutes, or sometimes tens of minutes.

A number of the reports indicate that combinations of two or more kinds of sound are heard; most frequent are reports of hissing noises followed by a sharper 'crackling.' Many hearers said that the sound was distinctly audible but not loud. If more than one person was present when sound was heard, most reports stated that all heard the sound, but other reports specifically noted that not all persons did. A few reports describe incidents wherein persons within a proximity of a few hundred meters—but not in communication—simultaneously heard aurorally related sounds. Four of the 100 reports in the Alaska collection also mention associated odors like that of ozone.

Most people who have reported hearing sound were watching the aurora at the time, and they typically stated that the sound and the aurora varied with some degree of synchronism. The statement that the sound was heard "with every rush of the aurora" is not atypical, and a few reports stated that the hearer observed a one-to-one correspondence between sounds and auroral variations.

Such observations are the basis for much of the skepticism voiced by scientists about the reality of auroral sound, and the suggestion that the 'hearing' is a purely psychological phenomenon. For some time, scientists have been aware that if the aurora generates sound at auroral altitude—roughly 80 to 100 km—the sound waves will require approximately 5 min to travel to the earth's surface. This long travel time certainly is inconsistent with the reports of simultaneous sound and auroral variation. Furthermore, it is now known that the energy deposited in the upper atmosphere by auroral processes is insufficient to produce sound waves in the audible range that can propagate to the earth's surface and be heard. Sound waves in the audible range are highly attenuated during travel through the atmosphere, although infrasonic waves of very long period, of order 10 sec, are generated by auroral processes and do propagate to the earth where they can be instrumentally recorded. Such waves cannot be detected by humans.

The impossibility of a person on the ground being able to hear any sound waves generated at auroral altitudes—and certainly not at the times the waves are produced—leads to the suggestion that the sensation of aurorally related sound is psychological; as one observer put it, the eye fooling the ear, or the instinctive feeling that anything that moves fast must make a noise.

Another explanation frequently offered is that people interpret other natural noises as being aurorally associated because the noises are heard when the aurora is highly active, but are ignored when the aurora is not. The human mind does indeed tend to focus on what it considers relevant, and to ignore what seems irrelevant. Also, the excitement that comes from watching a truly spectacular aurora may engender a heightening of the senses that leads to a sharpened awareness of everything in a person's surroundings, including sounds that otherwise might go unheard or be ignored. No doubt some of the reports of auroral sound are due to such mistaken identifications, but these reports appear to be in the minority.

From their analysis of the 200 individual reports contained in their paper on auroral sound, Silverman and Tuan concluded that the preponderance of the anecdotal evidence is that aurorally associated sound is a real, physiological phenomenon. An independent review of those reports and an analysis of the 100 other reports collected in Alaska brings me to the same conclusion. A factor strongly influencing the identical conclusions is the high degree of consistency in the anecdotal reports, both with respect to the nature of the sound and to the degree of association with certain specific auroral behavior and its location relative to the observer. Also, some of the reports strictly contradict a psychological explanation: those 5 to 10 percent that contain statements that the sound was heard before the aurora was seen, or that the sound could be heard when the eyes were covered.

Some 60 of the reports give enough information on both the sound and the aurora to permit the conclusion that the sound is most often or always heard when bright, rapidly moving discrete auroras are in or near the hearer's zenith area. Spirals (surges) with fast-moving ray structures are specified in some reports. A very few observers reported ozone smells, and fewer yet suggest near-ground lights that might be akin to St. Elmo's fire.

My analysis of the reports suggests that most or all of the reports relating to synchronism of the sound with auroral motions are compatible with a general rather than a one-to-one correspondence between sound and motions. People tend mentally to bring into unison two types of variation, one seen and one heard, that in fact are not quite synchronized. For example, if lively music is played when television recordings of rapidly varying auroras are viewed, the music and the aurora seem to be orchestrated together. It matters not exactly what music is played or when it is started; the human mind brings the musical and visual variations into a surprising degree of unison. Assuming the correctness of this conjecture, a mechanism to produce auroral sound need only be synchronized with auroral motions to within a few or a few tens of seconds.

Section 17.2
POSSIBLE MECHANISMS FOR AURORALLY ASSOCIATED SOUND

Assuming the reality of aurorally associated sound, the anecdotal reports set several requirements on any mechanism that might produce the sensation of sound. The sound cannot be the result of sound waves produced at auroral altitudes and propagated to the ground. The responsible mechanism must generate sound at the earth's surface or directly induce the sensation of sound in people's heads within a few or few tens of seconds of the time that brilliant, active discrete auroras swirl overhead the hearers. Although a few people have reported hearing aurorally associated sound numerous times, the sound appears to be relatively rare, so it is expected that the mechanism operates only under extreme circumstances, which may include an unusual electrical characteristic of the lower atmosphere. The mechanism must be capable of operating for times that range from a few seconds to several tens of minutes and to create sounds or the sensation of sounds that vary in intensity and have strong components in the frequency range 5 kHz to 15 kHz. If the few (approximately 10) reports of associated ozone-like smells are to be believed, then the mechanism should explain these. The mechanism must be able to operate without requiring the presence of manmade objects, such as buildings or other structures, since aurorally associated sound is reported from many locations, usually rural, where no such objects are nearby. The mechanism must produce a sensation of sound distinct enough to be sensed by most, but perhaps not all, persons appropriately located, even those who do not claim good hearing.

The aurorally associated sound appears to be similar to or identical with the so-called 'anomalous' sounds sometimes reported during the passage of extremely bright meteor fireballs. Unlike the sounds of detonations and shock waves heard later, the anomalous meteor sounds are simultaneous with the meteor's passage high overhead. The sounds are of two types: a variable hissing or swishing noise heard by observers located anywhere directly beneath the meteor's luminous trajectory, and sharper crackling or sputtering noises reported only by persons located directly under the point at which the fireball explodes. Rarely reported are associated smells, like that of onions or sulphur, which may be due to ozone. Anomalous meteor sound and aurorally associated sound may have a common mechanism, and several have been proposed as possible explanations for one or the other. Suggestions include very-low-frequency (VLF) radio waves produced by a high-altitude source and propagated to the ground, and electrical discharges produced at the ground.

Sec. 17.2.1 The VLF Electromagnetic Wave Mechanism

One investigator has proposed that the anomalous meteor sound might be due to VLF emissions produced in the turbulent wake of a bright meteor. Such waves will immediately propagate to the ground where they somehow might become transduced to sound waves. A possible prospect is a transduction by means of the piezoelectric effect within naturally occurring materials in the earth's surface. Piezoelectric materials can generate electrical voltages, perhaps strong enough to create discharges in the air. In fact, piezoelectric materials are sometimes used to start gas appliances; a sharp blow to the material produces a spark in the surrounding air that ignites the gas. The reverse also occurs: a piezoelectric subjected to an electrical voltage expands or contracts, and that motion might generate sound waves. The electric field in an electromagnetic VLF wave could conceivably cause enough motion in piezoelectric rocks and soil to produce audible noise.

It is also suggested that the electrostatic component of VLF electromagnetic waves possibly generated by turbulent meteor wakes might be perceived because they vibrate metal or poorly conducting material very close to the ear. Similarly, the vibration might be on the surface or inside the ear in a way that produces the sensation of sound without the entry of a sound wave as such. Laboratory tests using known electrostatic fields varied with frequencies in the auditory range have shown that humans can detect "sounds" in this way if exposed to peak-to-peak variations of 160 volts per meter (V/m) or greater. The sensitivity varies much between individuals.

This mechanism—generation of VLF electromagnetic waves at a high-altitude source, their rapid propagation to the ground at the speed of light, and then their production of sound waves near a hearer or the sensation of sound within the hearer—has possibilities for aurorally associated sound as well, because processes accompanying the aurora do produce VLF waves. Measurements on satellites and on the ground show that the generation takes place in the lower magnetosphere, but somewhat above the aurora. Called auroral hiss, the generated waves are strongest when bright, active discrete auroras are in evidence—the kind seen when auroral sound is heard. However, the strength of the signals measured at the ground is approximately 100 times less than that required for direct sensory perception. Thus, if aurorally associated VLF waves are the cause of aurorally related sound, the waves must somehow get transduced to actual sound waves in the vicinity of those who hear auroral sound. The VLF waves do carry enough energy that a conversion of less than 1 percent would be sufficient to generate audible sound waves.

Sec. 17.2.2 The Coronal Discharge Mechanism

More than 30 years ago, the well-known Norwegian auroral scientist Carl Störmer suggested that the probable cause of auroral sound was coronal discharge in the vicinity of hearers. He conjectured that electric fields associated with the aurora caused the coronal discharge.

Coronal discharge is the most common name applied to the breakdown of air—its ionization—when subjected to a very high electric field. Earlier in this book, 'coronal' was left off the term when discussing lightning strokes and the spark that appears across the gap of a spark plug in an automobile engine. Coronal discharge is a familiar phenomenon that occurs naturally in the atmosphere; lightning perhaps is the most cited example. In dry air, especially after one person has collected electrical charge by shuffling across a rug, coronal discharge can make kissing an especially exhilarating experience. Coronal discharge also frequently occurs on high-voltage transmission lines, causing a part of the normal energy loss in electrical transmission. Coronal discharge seems capable of producing all of the observed phenomena associated with reports of aurorally associated and anomalous meteor noise because, even before substantial current appears in a discharge, it emits audible hissing and crackling noises of variable intensity, VLF electromagnetic noise, and ozone. The discharge typically appears near sharply pointed conducting objects where high curvature intensifies an electric field enough to cause air breakdown. The required electric field at the point of initiation is extremely high, some millions of volts per meter. Despite the high voltage needed there, measurements have shown that coronal discharge will take place on pointed objects when the electric field observed a meter or so away is only a few thousand volts per meter. One investigative group observed the initiation of discharge on a nearby spruce tree when the measured field was 6000 V/m, and on a close-by sharp metal point, the discharge began at 2000 V/m. Fields of that intensity often occur near the earth's surface underneath thunderclouds, although the average vertical electric field intensity during periods of fair weather is but 130 V/m.

If coronal discharge is the cause of aurorally associated sound, auroral phenomena must somehow create near ground level the high electric fields required. Unfortunately, mathematical analysis of the situation is complex, and scientists so far have been unable to obtain a satisfactory quantitative understanding even of how thunderstorm clouds become electrified enough to create coronal discharges. They do know, however, that the transport of electrical charge in the lower atmosphere is very sluggish. This sluggishness allows intense concentrations of charge to build up in the air or on ungrounded objects so that powerful electric fields result.

Too few measurements of the atmospheric electric field have been made in conjunction with auroras. Those few observations have revealed that discrete auroras can occur overhead with absolutely no effect on the electric field measured at the ground. On the other hand, measurements on the ground and on balloons flown beneath auroral displays have revealed aurorally associated variations, over the course of seconds or minutes, as great or more than 1000 V/m. Such variations are at least in the ball park where coronal discharge can occur.

An intuitive argument can be proposed that suggests how the comparatively small electric fields associated with fast-moving discrete auroras (fields probably not often exceeding 1 V/m) can unbalance the atmospheric column below in a way that builds a high electric field near the ground, perhaps high enough to produce coronal discharge. In part, the argument involves the occurrence of abnormally low electrical conductivity in the air column such as might occur when the air is extremely free of water vapor and solid particulates. Thus, the idea is compatible with the observations that not all fast-moving auroras produce auroral sound; that is, the condition of the atmosphere may be an additional determining factor.

Even if the electric fields associated with the aurora in the ionosphere are incapable of developing enhanced electric fields at the ground sufficient to produce coronal discharge, one other possibility comes to mind. The normal electric field at the ground is near 130 V/m, and it is claimed that some people can directly sense electric field variations of as little as 160 V/m. If in some way the aurorally associated electric field in the ionosphere can cause rapid variations in an only slightly greater than normal "fairweather" field, then people might directly sense the variations and thereby "hear" auroral sound. If the variation—even a slow one—becomes substantially greater, it can also produce coronal discharge. Thus we are left with the interesting possibility that aurorally associated sound might be due to two causes: the direct sensing of variable electrostatic fields too weak to create coronal discharge, and the actual hearing of sound produced in nearby objects by coronal discharge.

The mystery of auroral sound remains unanswered. Nor is it easy to design instrumental measurements that will provide a final solution. In 1962, Dr. Eugene Wescott of the Geophysical Institute began a two-year observational program on auroral sound at Fairbanks, Alaska. He used sensitive but highly directional microphones pointed up toward the sky, and obtained only inconclusive results. He and others now agree that this procedure was unlikely to detect sounds generated near the ground by coronal discharge or other mechanisms. A long-term program that includes the taking of electric field measurements would seem to be in order, and since the conditions that produce auroral sound appear to be rare, the effort might take years before a definitive answer is obtained. In the meantime, all we can do is watch, listen and wonder.

Auroral Sound
Relevant Journal Articles

Ingalls, C. E., Sensation of hearing in electromagnetic fields, *New York J. of Medicine, 67*, 2992-2997, 1967.

Hutchinson, W. C. A., and I. M. Stromberg, New pulse technique for measuring point discharge in the atmosphere, *Nature, 222,* 654-655, 1969.

Olson, D. E., Auroral effects on atmospheric electricity, *Pure and Applied Physics (Pageoph),* 84, 118-139, 1971.

Silverman, S. M., and T. F. Tuan, Auroral audibility, *Advances in Geophysics,* 16, 155-266, 1973.

Markson, Ralph, Solar modulation of atmospheric electrification and possible implications for the Sun-weather relationship, *Nature, 273*, 103-109, 11 May 1978.

Kelley, M. C., an article on electrostatic waves from auroral field lines, pages 53-71,; also, Gurnett, D. A., An article on electromagnetic waves from auroral field lines, pages 91-106, in Auroral Processes (Advances in Earth and Planetary Sciences 4, Special Issue of *Journal of Geomagnetism and Electricity*, C. T. Russell, Ed., 1979.

Keay, C. S. L., Anomalous sounds from the entry of meteor fireballs, *Science, 210,* 11-15, 3 October 1980.

GLOSSARY

ALL-SKY CAMERA. A camera that photographs the entire sky by employing a fisheye lens or a spherical mirror system.

ANGSTROM UNIT. A unit of length equal to 10^{-8} centimeters or 1/10th nanometer (a nanometer is one billionth of a meter).

ARC. An auroral form usually extended magnetic east-west and with uniform curvature.

ARTIFICIAL AURORA. A man-made aurora.

ATOM. The smallest particles of an element, consisting of electrons revolving around a nucleus composed of protons and neutrons. An atom contains equal numbers of electrons and protons.

AURORA. Also called the northern lights, aurora polaris, and various other names; a visible light emitted from the gases in the high atmosphere when struck by fast-moving particles coming in from above. The aurora is highly variable and it results, along with various related phenomena, from solar energy carried it the upper atmosphere by a complex series of processes.

AURORAL CORONAL POINT. The point in the sky toward which auroral rays appear to converge. It is located in the magnetic zenith.

AURORAL DISPLAY. A collection of auroral forms seen at one time or over some period of time such as one night.

AURORAL ELECTROJET. Intense electric current in the auroral ionosphere, typically eastward in evening and westward in morning.

AURORAL FORM. An individual auroral structure visible to the eye.

AURORAL OVAL. The annular-like belt around the geomagnetic pole in which various auroral forms lie at a particular moment.

AURORA POLARIS. Name for the aurora of both the northern hemisphere and the southern. Aurora borealis refers to that in the northern, and aurora australis to that in the south. For other names see Section 14.1.

AURORAL SPIRAL. A highly convoluted auroral form that takes on a vortex shape, similar to but much larger than a *curl*.

AURORAL SUBSTORM. A global burst-like increase in auroral activity lasting one to several hours, usually commencing on the midnight meridian and expanding both east and west along the auroral oval. Several occur each day; they are more frequent and larger during periods of high activity.

AURORAL ZONE. Circular belt around the geomagnetic pole in which the frequency of seeing auroral forms is greatest over a long period of time. The peak of the auroral zone is a line defining the maximum seeing frequency within the auroral zone.

BAND. Similar to an arc but with irregular and more extensive curvature.

BAND EMISSION. Light emitted from molecules in closely spaced and structurally related emission lines.

BAR MAGNET. A magnet possessing the simplest possible (dipole) magnetic field like that of a current loop or a magnetized sphere.

BLACK AURORA. Pronounced voids in a widespread diffuse aurora that appear unusually dark compared to other regions of the sky.

BREAK-UP. The rapid increase in auroral activity seen by an observer during the early expansive phase of an auroral substorm.

COLLISION. Energy-transferring interaction between two particles, as when a fast electron strikes an atom or molecule, or when two atoms or molecules collide.

CONJUGACY. Referring to the aurora, the term means the degree of similarity of auroras at conjugate points.

CONJUGATE POINTS. Points located on the same geomagnetic field line, usually points at the same altitude, one in the northern hemisphere and one in the southern.

CORONA. The auroral form seen when rayed aurora surrounds the observer's magnetic zenith.

CORONAL HOLE. Dark region on the sun from which intense solar flares emanate to contribute to high solar wind speeds, more so perhaps than those rising from near sunspots.

CURL. A vortex structure that forms an auroral ray. A curl can be recognized only when it appears in the magnetic zenith.

DIPOLE MAGNETIC FIELD. The magnetic field of a current loop, also of a bar magnet, and of a magnetized sphere. The earth's magnetic field is almost dipolar.

DISCRETE AURORA. Aurora with well-defined boundaries, variable brightness, and always aligned upward in the direction of the local magnetic field.

DIFFUSE AURORA. Weak, poorly defined aurora, usually without well defined boundaries. It can occur in horizontal sheets. See also Pulsating Aurora.

D REGION. The lowest layer in the ionosphere, at about 60 km.

ELECTRICITY. A property of certain fundamental particles of matter, as electrons and protons, that have a force field associated with them and that can be separated by the expenditure of energy.

ELECTRON. An elementary negatively charged particle carrying unit charge that forms a part of all atoms.

EMISSION. In this book, light emitted from an excited atom or molecule.

ENERGY. $E = mc^2$. It's what makes everything work, and we think of it as coming in many forms. The energy of motion, we call kinetic energy, the energy of position, potential energy (the higher we go up a flight of stairs, the more potential energy we possess). Energy locks electrons in atoms, and when extra energy is added to an atom—in only set allowable amounts—the atom becomes excited and can give off the energy in the form of photons, producing the light we see. Energy is locked up in force fields: the earth's gravity field, its magnetic field and all the stray electric fields that roam about because they are created by concentrations of charged particles that seem to crop up here and there.

ENHANCED AURORA. Auroral forms that contain one or more bands of enhanced brightness near their lower borders.

E REGION. A middle layer of the ionosphere, at about 100 km.

EXCITE. To add energy to an atom or molecule, raising it to an excited energy state.

EXCITATION. The delivery of sufficient energy to an atom or molecule to raise it to an exited state where it may emit the energy in the form of a photon.

FAST AURORAL WAVES. Rapid, equatorward-moving enhancements of pre-existing auroral forms that give the appearance of a searchlight sweeping across the sky.

FLAMING AURORA. Aurora that contains waves of light enhancement rising from their lower borders.

FLICKERING AURORA. Rarely seen bright aurora containing ray-like striations that appear to oscillate like the flame of a candle in a draft.

F REGION. The highest layer of the ionosphere, extending upward from about 150 km.

GEOGRAPHIC ZENITH. The point on the celestial sphere directly overhead the observer; also called the true zenith.

GEOMAGNETIC FIELD. The earth's magnetic field. It extends outward beyond the earth's surface to the outer boundary of the magnetosphere.

GEOMAGNETIC POLE. Point where the axis of the earth's dipole magnetic field reaches the earth's surface, also called a dipole pole.

GREEN LINE (Auroral). The brightest emission in the normal aurora, at 5577 A. It is produced by oxygen atoms.

GROUND STATE. The lowest energy state that an atom can occupy.

HOMOGENEOUS. Auroral forms that do not show internal striations or other structure.

HYDROGEN ARC. A weak, diffuse arc seen equatorward of the discrete part of an auroral display during the evening hours. It is rich in hydrogen emissions and is caused mainly by incoming protons. Also called the proton arc.

ION. As used here, an atom or group of atoms lacking an electron, hence it is positively charged. General use of the term allows an ion to be an atom or group of atoms which contain an excess electron and therefore are negatively charged.

IONIZE. To remove an electron from an atom or molecule, changing it to a positive ion.

IONOSPHERE. The region in the high atmosphere, extending from about 60 km upward to several hundred km, that contains appreciable charged particles (electrons and positive ions). It reflects radio waves of certain frequencies.

LIFETIME (of an exited state). The time taken for 50 percent of the atoms or molecules in an excited state to emit photons and thereby fall to a lower energy state. In the physics of nuclear reactions, the lifetime is called the half-life.

LINE EMISSION. Light emission at a specific wavelength when an excited atom emits a photon to fall to a state of lower energy, as the oxygen green line at 5577 A.

LOWER BORDER. The bottom edge of an auroral form. It generally lies parallel to the earth's surface.

MAGNETIC BOTTLE. A magnetic field configuration wherein the field converges in two regions so that it can trap charged particles indefinitely. The Van Allen belt is a magnetic bottle.

MAGNETIC FIELD LINE. An imaginary line lying along the direction of the magnetic field. The directions of the field near a magnet can be seen by sprinkling iron filings on a piece of paper near the magnet. The filings align along the 'field lines.'

MAGNETIC POLE. Point on the earth's surface where the magnetic field is exactly vertical.

MAGNETIC RECONNECTION (Merging). Process where two magnetic fields join, as across the boundary of the magnetosphere where geomagnetic field lines can hook onto those of the solar wind.

MAGNETIC STORM. Period lasting one to several days during which global changes to the earth's magnetic field occur. The main feature of a magnetic storm is depression of the earth's magnetic field at low and middle latitudes, the consequence of increased energy content of the Van Allen belt.

MAGNETIC ZENITH. The point on the celestial sphere seen when looking exactly in the direction of the local magnetic field.

MAGNETOHYDRODYNAMIC DYNAMO. A generator of electromotive force created by the movement of charged particles across a magnetic field, as at the boundary of the magnetosphere.

MAGNETOSPHERE. The region above the ionosphere wherein the geomagnetic field controls the motion of charged particles. It extends outward to form a comet-like cavity in the solar wind. Within the outer magnetosphere, the geomagnetic field is highly distorted but in the inner magnetosphere the geomagnetic field is essentially dipolar in shape.

MAGNETOSPHERE BOUNDARY (Magnetopause). The outer boundary of the magnetosphere that separates it from the solar wind.

MIDNIGHT MERIDIAN. The meridian on which local midnight occurs, 180 degrees of longitude removed from the noon meridian.

The dawn-dusk meridian is 90 degrees of longitude removed from the noon and midnight meridians.

MIDNIGHT SECTOR. The region near the midnight meridian.

MIRROR POINT. The point at which a charged particle travelling in a converging magnetic field reverses its direction of forward motion (mirrors).

MOLECULE. Usually composed of two or more atoms of like kind (as the oxygen molecule O_2) or atoms of different kinds, such as salt (NaCl).

NORTHERN LIGHTS. The aurora, a persistent glow from gases in the high-latitude atmosphere when struck by incoming charged particles, mostly electrons.

PHOTOMETER. Device for measuring light, perhaps at specific wavelengths or over some range of wavelength.

PHOTON. A quantum of light energy that has both particle and wave characteristics. It has momentum but not mass, and can be identified by its wavelength.

PIEZOELECTRIC. A material that changes shape when an electrical voltage is applied across it, or which will generate a voltage if caused to change shape.

PLASMA. Matter partly or entirely ionized, as in the ionosphere, the magnetosphere, the solar wind and the sun; also in the deep interior of the earth.

PRIMARY ELECTRON (or Proton). Electron (or proton) that streams down into the atmosphere and that usually carries substantial kinetic energy. Primaries may strike atoms and molecules and eject secondary electrons from them.

PROTON. An elementary particle that carries positive unit charge and is found in all atomic nuclei. It is 1836 times heavier than an electron. One proton and one electron form a hydrogen atom.

PULSATING AURORA. Weak, usually diffuse aurora that exhibits rapid changes in brightness so that it appears to blink on and off. Sometimes also called diffuse aurora.

QUANTUM. An indivisible quantity of energy, like the energy carried by a photon.

RAY. A convolution in an arc or band that extends upward exactly in the direction of the local magnetic field. Seen in the magnetic zenith, a ray appears as a curl, a small vortex structure.

RAYED AURORAS. Those that contain rays.

RED LINE (Auroral). Name generally applied to two closely spaced line emissions from oxygen at 6300 and 6364 A, the brightest emission in the all-red aurora that occurs during highly disturbed periods.

RESONANCE SCATTERING. Absorption and re-emission of light at a particular wavelength.

SAINT ELMO'S FIRE. A visible electrical discharge from pointed objects such as trees or the tips of masts.

SECONDARY ELECTRON. An electron ejected from an atom or molecule by an incoming primary. The impact typically gives the secondary electron enough kinetic energy that it can move on into the atmosphere to cause other ionizations and also excitation of resident atoms and molecules.

SOLAR FLARE. A bright prominence on the sun which tends to cast out material from the sun and contribute to high solar wind speeds. Flares generally are associated with sunspot groups or with dark areas called coronal holes.

SOLAR WIND. Solar particles, mainly electrons and protons, that flow out from the sun with a supersonic speed and reach to the outer fringe of the solar system.

SUNLIT AURORA. Usually tall, rayed aurora blue-purple in color owing to resonance scattering of sunlight from ionized nitrogen molecules.

SUNSPOT. Dark region on the sun with which solar flares may be associated. The number of sunspots observed each year is a measure of solar activity.

TRANSITION. The movement of an excited atom or molecule to a lower energy state by emission of a photon.

VAN ALLEN BELT. A trapping region (magnetic bottle) in the inner magnetosphere that contains variable populations of charged particles. It is sometimes referred to as the Van Allen belts because it contains two identifiable regions.

VEIL. A broad expanse of auroral light, usually of uniform brightness.

SUGGESTED FURTHER READING

GENERAL READERSHIP BOOKS

S.-I Akasofu, *Aurora Borealis*, The Amazing Northern Lights (Anchorage, Alaska: The Alaska Geographic Society, Quarterly, Vol. 6, No. 2, 1979).....{Authoritative, contains a reprint of a nice article on auroral legends by Dorothy Jean Ray, *Alaska Sportsman*, April 1958.}

Robert H. Eather, *Majestic Lights*, The Aurora in Science, History and the Arts (Washington: American Geophysical Union, 1980).....{Comprehensive in history and the arts; a generally excellent book.}

Asgeir Brekke and Alv Egeland, *The Northern Light*, From Mythology to Space Research [Translation from Norwegian] (Berlin: Springer-Verlag, 1983).....{Authoritative, features Scandinavian auroral science.}

Harald Falck-Yetter, *Aurora*, The Northern Lights in Mythology, History and Science [Translation from German] (Spring Valley, New York: Anthroposophic Press, 1983).....{Features mythology and history; beware some of the scientific explanations.}

SELECTED TECHNICAL BOOKS

J. A. Ratcliffe, *An Introduction to the Ionosphere and Magnetosphere* (Cambridge: At the University Press, 1972).....{Compact; more readable than most highly technical books; a useful general reference.}

Carl Störmer, *The Polar Aurora* (Oxford: At the Clarendon Press, 1955).....{A classic, contains detailed information on auroral height determinations in Norway.}

S.-I. Akasofu, *Physics of Magnetospheric Supstorms* (Dordrecht, Netherlands: D. Reidel Publ., 1977.....{Contains descriptive coverage and interpretation of many aurorally related phenomena.}

Sydney Chapman and Julius Bartels, *Geomagnetism, Volumes I and II* (Oxford: At The Clarendon Press, 1940).....{An old classic on the earth's magnetic field and its variations.}

Anders Omholt, *The Optical Aurora* (New York: Springer-Verlag, 1971).....{Emphasizes spectroscopic aspects of the aurora.}

A. Vallance-Jones, *Aurora* (Dordrecht, Netherlands: D. Riedel, 1974).....{Emphasizes spectroscopic aspects of aurora.}

READABLE ARTICLES IN RECENT PUBLICATIONS

S.-I. Akasofu, Aurora, *Encyclopedia of Physical Science and Technology*, [Academic Press] 1987, pp 309-320.....{Modern general discussion of auroral generation processes.}

S.-I. Akasofu, The Dynamic Aurora, *Scientific American,* May, 1989, pp 90-97.....{Emphasis on the solar wind-magnetosphere generator that powers auroral phenomena.}

T. J. Hallinan, Auroras, *Geomagnetism*, [Academic Press] Vol. 4, pp 742-798, 1991....{General, but emphasizes auroral morphology.}

Neil Davis, Chemical Releases in the Ionosphere, *Reports on Progress in Physics, Vol. 42, No. 9,* 1979, pp 1565-1604.....{Discussion of artificial auroras and chemical releases in the atmosphere.}

Neil Davis, The Aurora, *Geophysics: The Leading Edge of Exploration, Vol. 3, No. 9,* [Society of Exploration Engineers] September 1984, pp 34-41....{General discussion of aurora and its related effects of importance to exploration geophysicists.}

INDEX

D

E

F

R

S

T

U, V